drafting
for industry

Worksheets

by

WALTER C. BROWN

South Holland, Illinois

THE GOODHEART-WILLCOX COMPANY, INC.

Publishers

INTRODUCTION

The WORKSHEETS are designed for use with the text, DRAFTING FOR INDUSTRY. Its purpose is to:
1. Help to develop basic industrial drafting skills.
2. Make the study of drafting more interesting and productive for you by increasing the number of new learning activities and eliminating much of the repetitive work.
3. Provide an opportunity to utilize state-of-the-art drafting techniques and equipment.

The WORKSHEET problems are closely correlated with the text. They have been carefully selected from the drafting rooms of industry to provide realistic experiences. Solutions to the problems may be obtained using traditional or computer-based means. However, the WORKSHEETS for Chapter 31 and 32 are intended to be with a CAD system.

The text, DRAFTING FOR INDUSTRY, and the WORKSHEETS give you a better understanding of the industrial drafting field and its many career opportunities.

CONTENTS

Copyright 1990
by
THE GOODHEART-WILLCOX COMPANY, INC.

Previous Editions Copyright, 1984, 1980, 1974

1234567890-90-8765432109

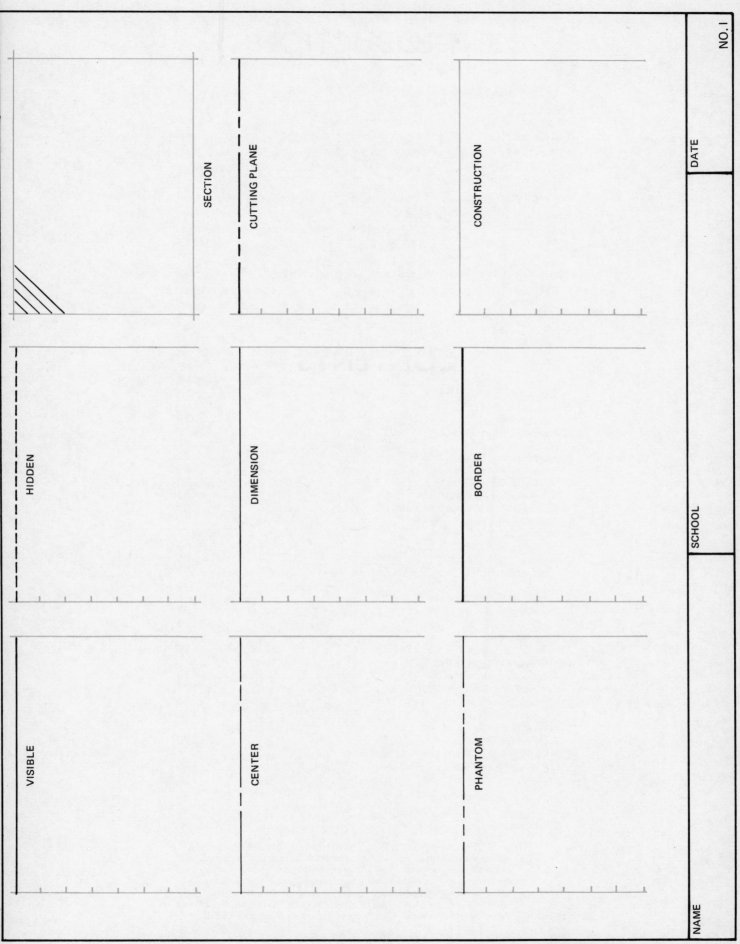

SECTION

CUTTING PLANE

CONSTRUCTION

HIDDEN

DIMENSION

BORDER

VISIBLE

CENTER

PHANTOM

DATE

SCHOOL

NAME

DATE

SCHOOL

SCALE	LENGTH OF LINE
$1'' = 1''$	$6\frac{3}{4}''$
FULL SIZE	$5\frac{1}{8}''$
$1'' = 1' - 0''$	$7' - 3\frac{1}{2}''$
$\frac{1}{4}'' = 1' - 0''$	$30' - 9''$
$\frac{1}{4}'' = 1' - 0''$	$22' - 5''$
$1\frac{1}{2}'' = 1' - 0''$	$3' - 7\frac{1}{2}''$
QUARTER SIZE	$2' - 2\frac{1}{4}''$
$\frac{1}{2}'' = 1' - 0''$	$14' - 11''$
$\frac{1}{8}'' = 1' - 0''$	$62' - 10''$
$1'' = 1' - 0''$	$6' - 4\frac{3}{4}''$
$\frac{3}{16}'' = 1' - 0''$	$40' - 6''$
$\frac{3}{32}'' = 1' - 0''$	$81' - 0''$

NAME

SCALE	LENGTH OF LINE
1" = 10" (USE 10 SCALE)	9.80"
1" = 10' - 0"	4.40'
1" = 30" (USE 30 SCALE)	226.00"
FULL SIZE (USE 50 SCALE)	6.375"
FULL SIZE	3.187"
FULL SIZE	7.250"
FULL SIZE	6.125"
FULL SIZE	4.020"
FULL SIZE	5.825"
1" = 500 MILES (USE 50 SCALE)	2475 MILES
1" = 500'	4250.5'
1" = 2" (USE 20 SCALE)	14.75"

NAME

SCHOOL

DATE

NO. 3

3

SCALE	LENGTH OF LINE
1 : 1 (USE 1 : 100 SCALE)	127 mm
1 : 1	98.5 mm
1 : 1	152.25 mm
1 : 1	106 mm
1 cm = 1 m (USE 1 : 100 SCALE)	17.54 m
1 cm = 1 m	13.08 m
1 cm = 1 km (USE 1 : 100 SCALE)	16.55 km
1 cm = 1 km	20.6 km
FULL SIZE (USE 1 : 100 SCALE)	105.2 mm
FULL SIZE	86.5 mm
HALF SIZE (USE 1 : 20 SCALE)	382 mm
HALF SIZE	264 mm

NAME

SCHOOL

DATE

NO. 4

DRAW LINES AS INDICATED.

NO. 5

HORIZONTAL

VERTICAL

45° TO THE RIGHT

30° TO THE LEFT

60° TO THE RIGHT

75° TO THE LEFT

75° TO THE RIGHT

15° TO THE LEFT

NAME	SCHOOL	TITLE: INSTRUMENT LINES	DATE
			NO. 5

DRAW LINES PARALLEL TO GIVEN LINE BY USE OF TRIANGLE AND T-SQUARE OR TWO TRIANGLES.

NO. 6

DRAW LINES PERPENDICULAR TO THE GIVEN LINES BY USE OF TRIANGLES.

LAY OFF THE REQUIRED LINES AS INDICATED BY USE OF A PROTRACTOR.

41° WITH HORIZONTAL

87° 30' WITH VERTICAL

DIVIDE THE LINES INTO EQUAL PARTS AS INDICATED BY USE OF DIVIDERS.

7 EQUAL PARTS

9 EQUAL PARTS

NAME	SCHOOL	TITLE: INSTRUMENT LINES	DATE
			NO. 6

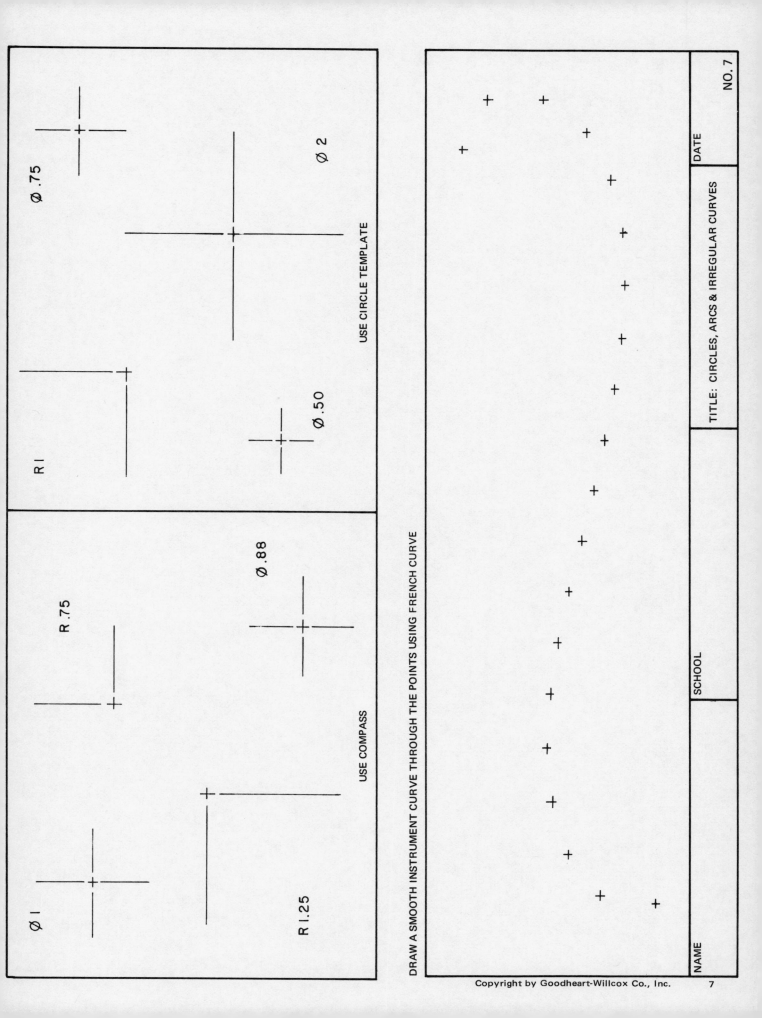

USE CIRCLE TEMPLATE

Ø .75

Ø 2

R 1

Ø .50

USE COMPASS

R .75

Ø .88

Ø 1

R 1.25

DRAW A SMOOTH INSTRUMENT CURVE THROUGH THE POINTS USING FRENCH CURVE

NAME

SCHOOL

TITLE: CIRCLES, ARCS & IRREGULAR CURVES

DATE

NO. 7

7

DRAW THE PARTS TO THE SCALE SHOWN. DO NOT DIMENSION.

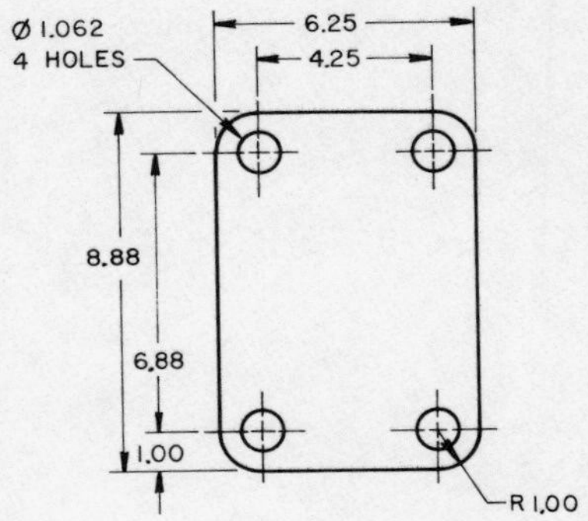

Ø 1.062
4 HOLES

6.25

4.25

8.88

6.88

1.00

R 1.00

SCALE : 1″ = 3″

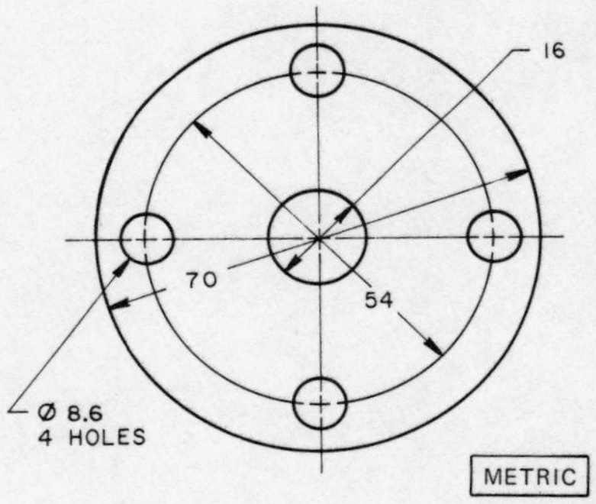

16

70

54

Ø 8.6
4 HOLES

METRIC

FULL SCALE

| NAME | SCHOOL | TITLE: INSTRUMENT DRAWINGS | DATE |
| | | | NO. 8 |

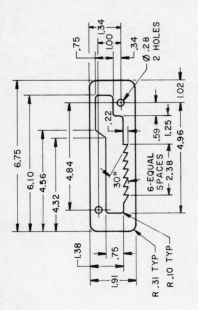

TITLE: GATE THROTTLE GUIDE
FULL SCALE

DATE

NO. 9

SCHOOL

NAME

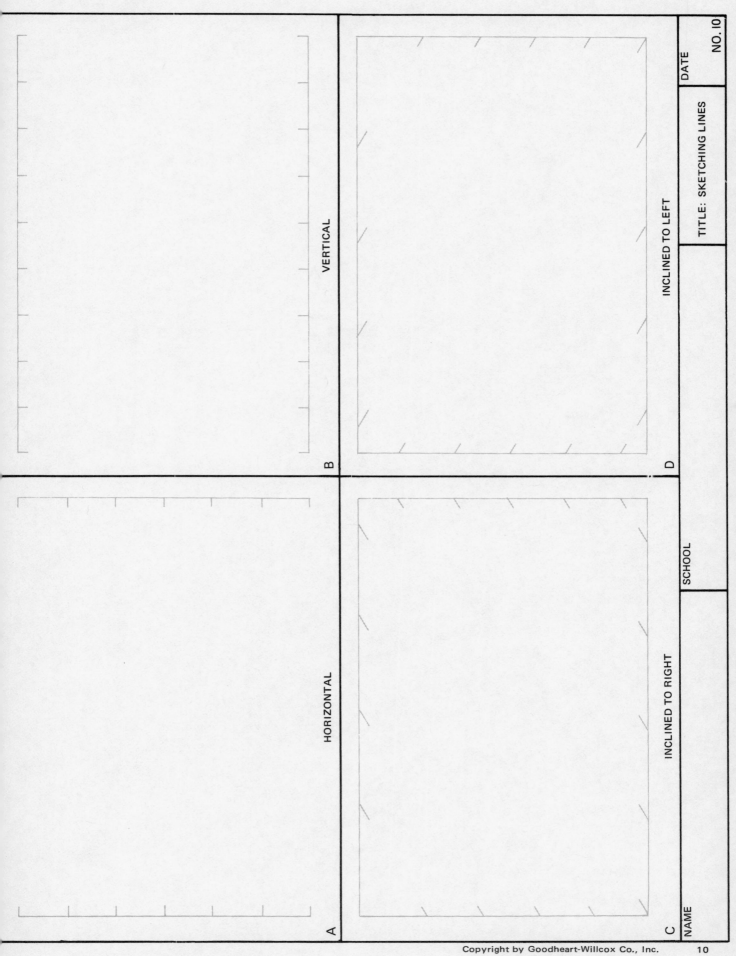

A

HORIZONTAL

B

VERTICAL

C

INCLINED TO RIGHT

D

INCLINED TO LEFT

NAME

SCHOOL

DATE

TITLE: SKETCHING LINES

NO. 10

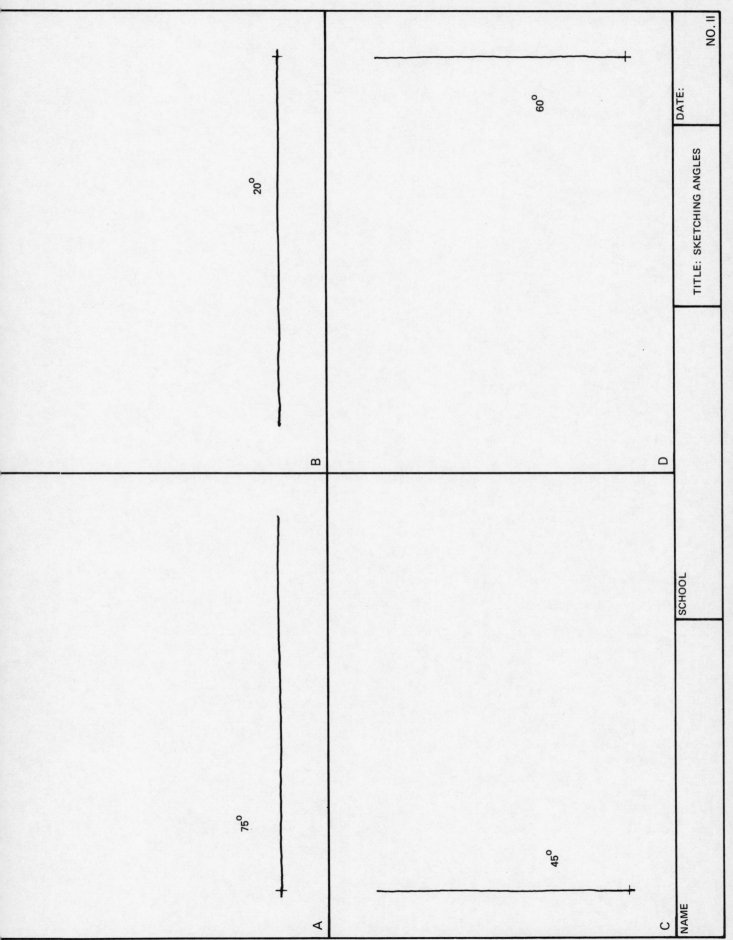

B 20°

A 75°

D 60°

C 45°

CIRCLE - ENCLOSING SQUARE METHOD

R 1

R .75

R 1.5

R .5

ARCS AS INDICATED

2.5" CIRCLE - CENTER LINE METHOD

2" CIRCLE - FREE CIRCLE METHOD

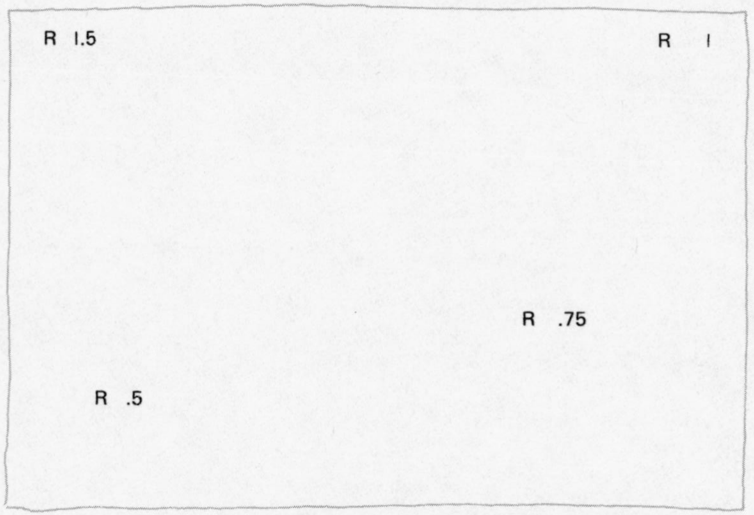

A

B

C

D

NAME

SCHOOL

DATE

TITLE: SKETCHING CIRCLES & ARCS

NO. 12

12

A

ELLIPSE - RECTANGULAR METHOD

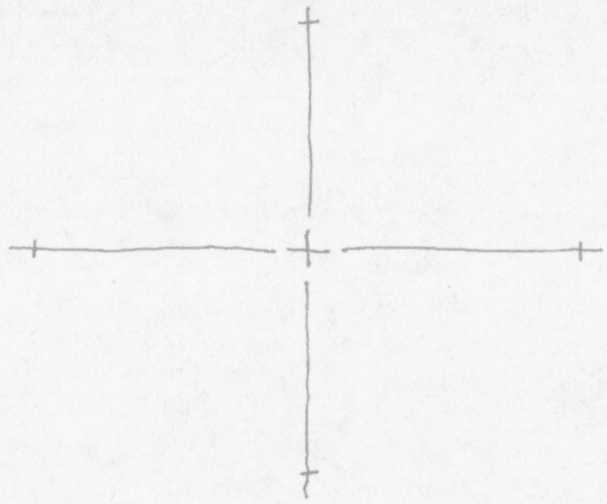

B

ELLIPSE - TRAMMEL METHOD

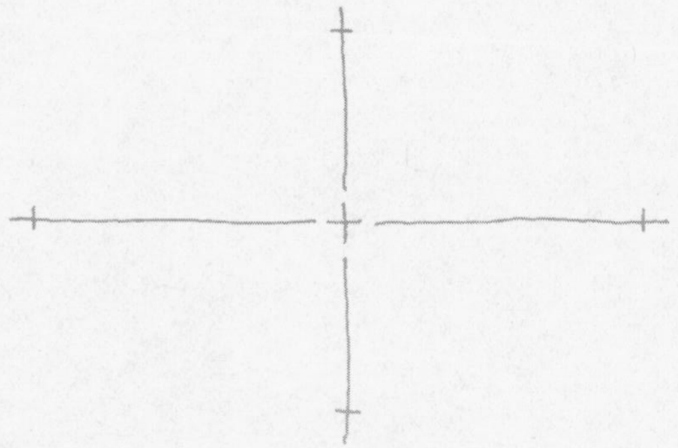

C

ELLIPSE - FREE ELLIPSE METHOD

D

IRREGULAR CURVE THRU POINTS

13

NAME

SCHOOL

TITLE: SKETCHING ELLIPSES & IRREGULAR CURVES

DATE

NO. 13

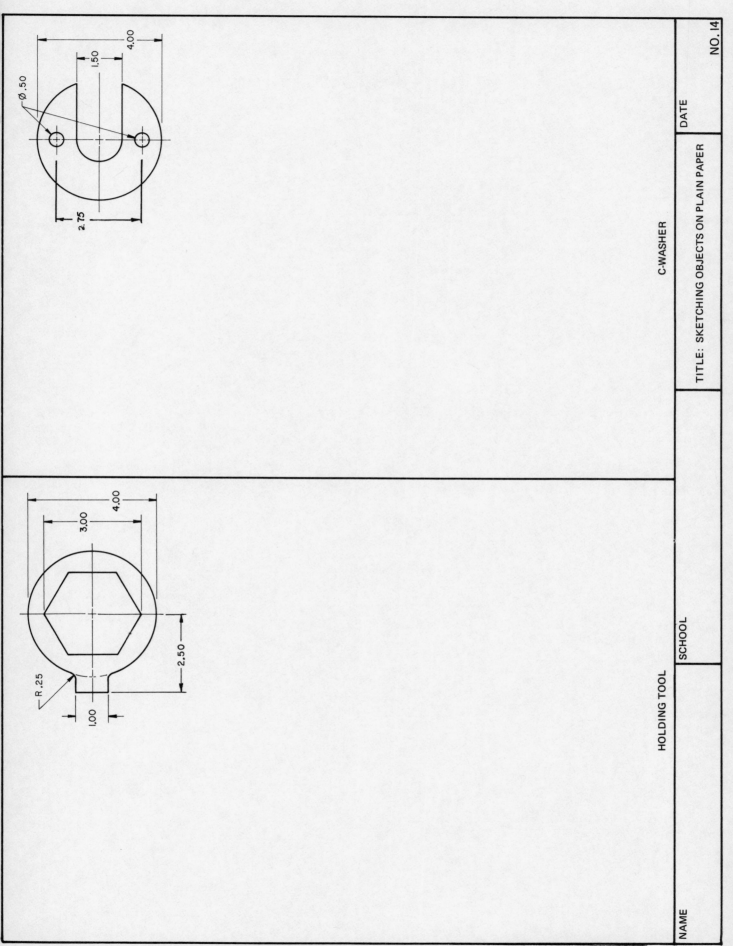

Ø .50

4.00

1.50

2.75

C-WASHER

4.00

3.00

R .25

2.50

1.00

HOLDING TOOL

TITLE: SKETCHING OBJECTS ON PLAIN PAPER

NAME

SCHOOL

DATE

NO. 14

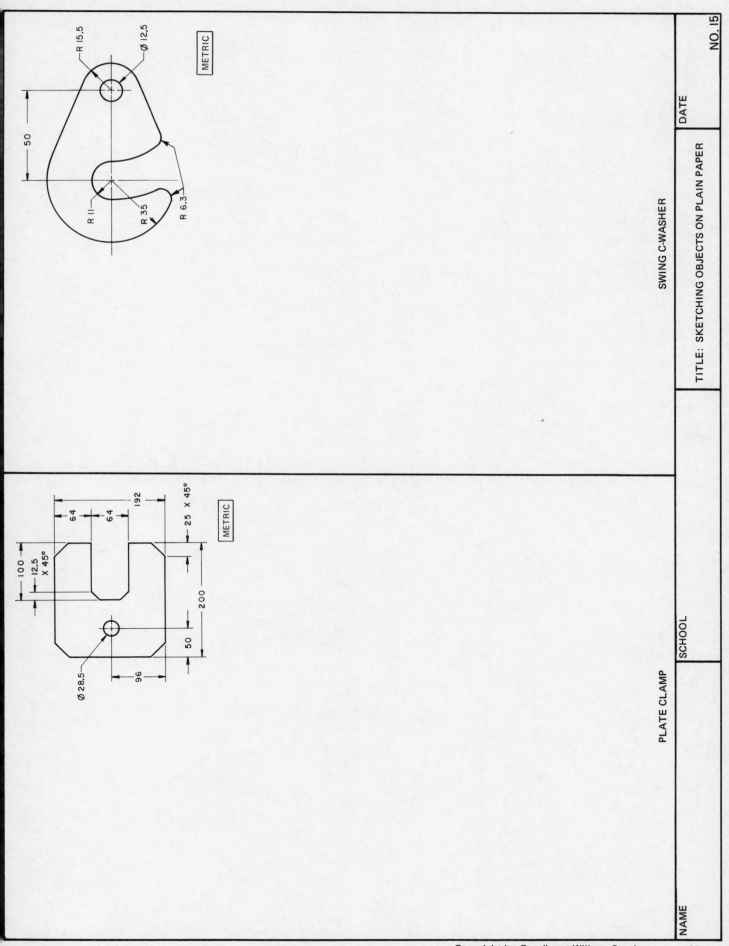

R 15.5

Ø 12.5

50

R 11

R 35

R 6.3

METRIC

SWING C-WASHER

192

64

64

100

12.5 X 45°

25 X 45°

200

50

96

Ø 28.5

METRIC

PLATE CLAMP

DATE

TITLE: SKETCHING OBJECTS ON PLAIN PAPER

NO. 15

SCHOOL

NAME

15

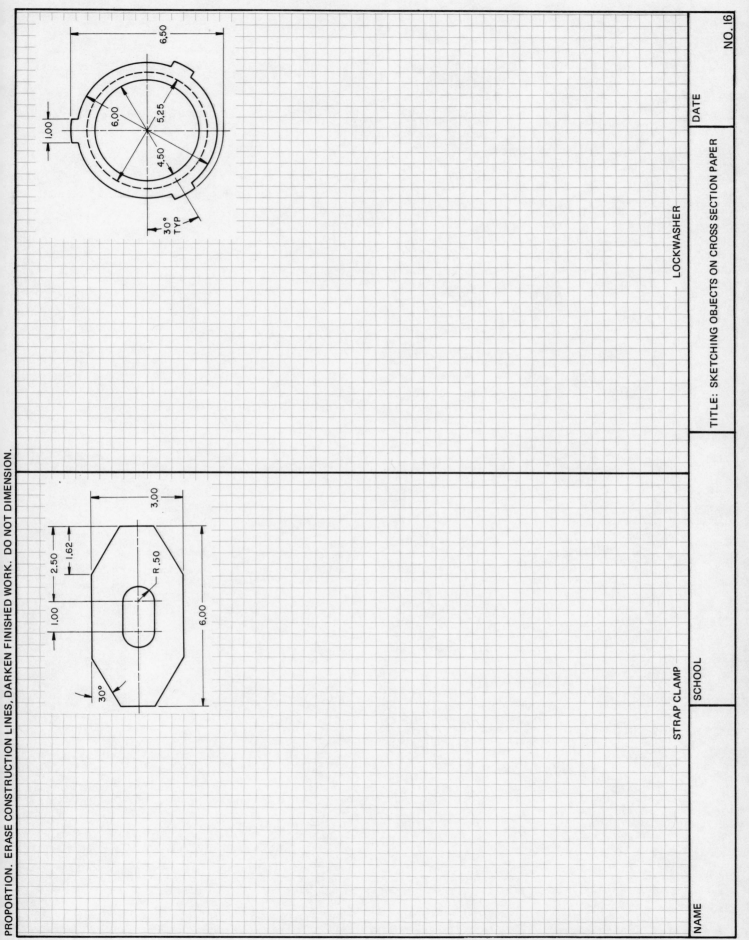

PROPORTION. ERASE CONSTRUCTION LINES, DARKEN FINISHED WORK. DO NOT DIMENSION.

6.50

6.00

5.25

4.50

1.00

30°
TYP

LOCKWASHER

3.00

2.50

1.62

1.00

R.50

6.00

30°

STRAP CLAMP

NAME

SCHOOL

TITLE: SKETCHING OBJECTS ON CROSS SECTION PAPER

DATE

NO. 16

16

PROPORTION. ERASE CONSTRUCTION LINES, DARKEN FINISHED WORK. DO NOT DIMENSION.

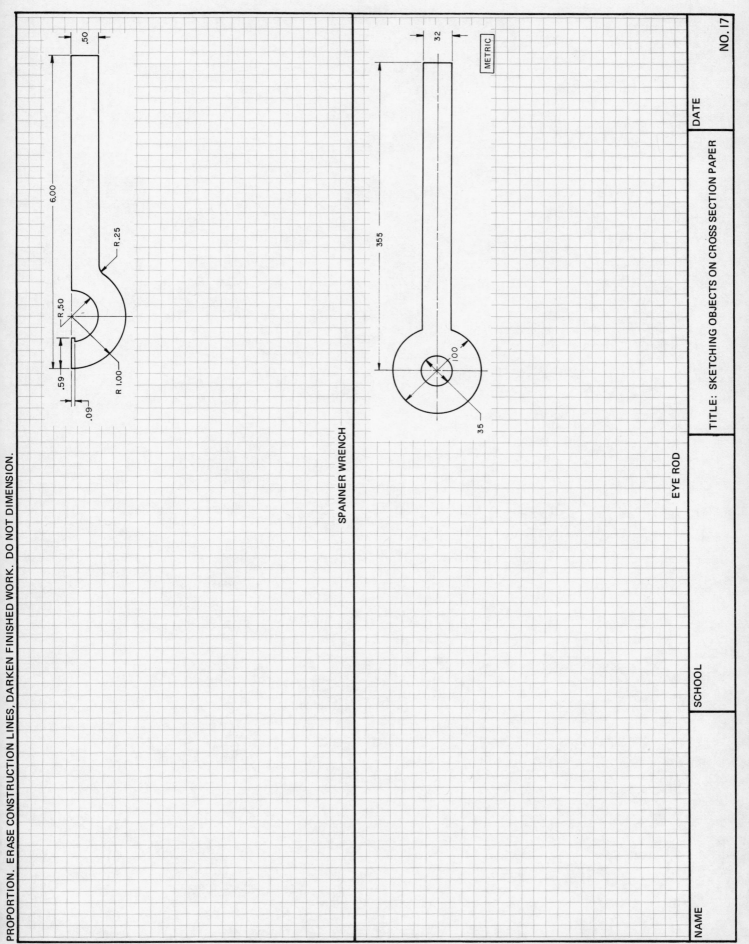

.50

6.00

R.25

R.50

R 1.00

.59

.09

SPANNER WRENCH

32

METRIC

355

1.00

35

EYE ROD

17

NAME

SCHOOL

TITLE: SKETCHING OBJECTS ON CROSS SECTION PAPER

DATE

NO. 17

SKETCH THOSE OBJECTS ASSIGNED BY YOUR INSTRUCTOR.

NAME:

SCHOOL

TITLE:

DATE:

NO. 18

18

ABCD

$\frac{1}{8}$" VERTICAL CAPITALS & NUMERALS

abcd

$\frac{1}{8}$" VERTICAL LOWER CASE LETTERS & NUMERALS

ABCD

$\frac{3}{8}$" VERTICAL CAPITALS & NUMERALS

TITLE: VERTICAL GOTHIC LETTERING

DATE

NO. 19

SCHOOL

NAME

19

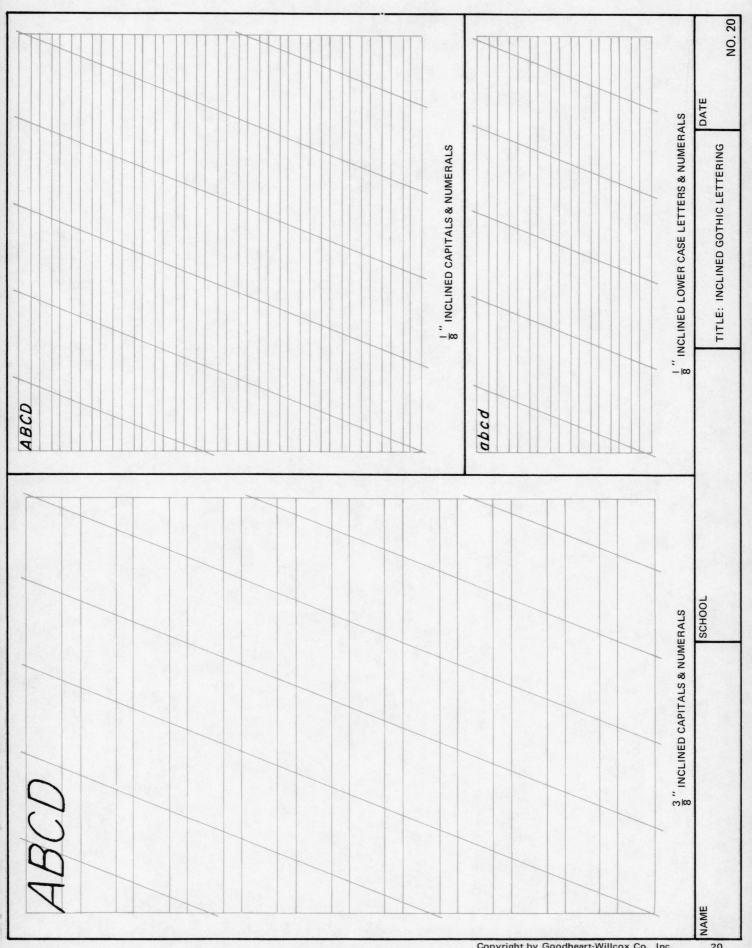

ABCD

3" INCLINED CAPITALS & NUMERALS
8

ABCD

1" INCLINED CAPITALS & NUMERALS
8

abcd

1" INCLINED LOWER CASE LETTERS & NUMERALS
8

TITLE: INCLINED GOTHIC LETTERING

NO. 20

NAME

SCHOOL

DATE

LETTER THE FOLLOWING NOTES IN THE SIZES INDICATED. USE VERTICAL OR INCLINED LETTERING AS DIRECTED BY YOUR INSTRUCTOR. STRIVE FOR UNIFORM AND NEAT LETTER FORMS. REFER TO THE TEXT AND LAY OUT AND LETTER THE TITLE BLOCK ACCORDING TO LAYOUT 1.

DIMENSIONS, TOLERANCES AND SURFACE FINISH VALUES APPLY BEFORE THE APPLICATION OF THE FINISHES

ROUTE MAIN HARNESS THROUGH EXISTING CLIP ON RIGHT SIDE OF PANEL ASSY, COWL TOP INNER FRONT

FLAME DEPOSIT UNION CARBIDE'S CHROME OXIDE COATING LC-4 .002-.004 FINISH THICKNESS ON NOTED SURFACE

FINISH AS FOLLOWS:
1. PAINT SYSTEM, ENAMEL, BLACK, M695476-II0, (FNSH NO. II0) SPERRY SPEC M6977I AS NOTED. NO OVER SPRAY PERMISSIBLE
2. PAINT SYSTEM, ENAMEL, WHITE, M695476-439, (FNSH NO. 439) SPERRY SPEC M692077 AS NOTED. NO OVER SPRAY PERMISSIBLE

SLOT MUST HAVE NO EXTERNAL BURRS

.203 DIA. HOLE - PIERCE FROM BOTH SIDES - FOUR PLACES

EMBOSS $\frac{5}{16}$ DIA X $\frac{1}{16}$ DEEP - INSIDE ONLY - FOUR PLACES

REMOVE BURRS AND SHARP EDGES

DIMENSIONS THROUGHOUT ARE TO 90° BEND AS ASSEMBLED ON PRODUCT. PART SHOULD BE OVERBENT TO 91° FOR ADDITIONAL TENSION

AN EASEMENT OF FOUR FEET ON EITHER SIDE OF THIS LINE IS RESERVED FOR UTILITY LINE THRU PROPERTIES

TITLE -- GEOMETRIC CONSTRUCTIONS.

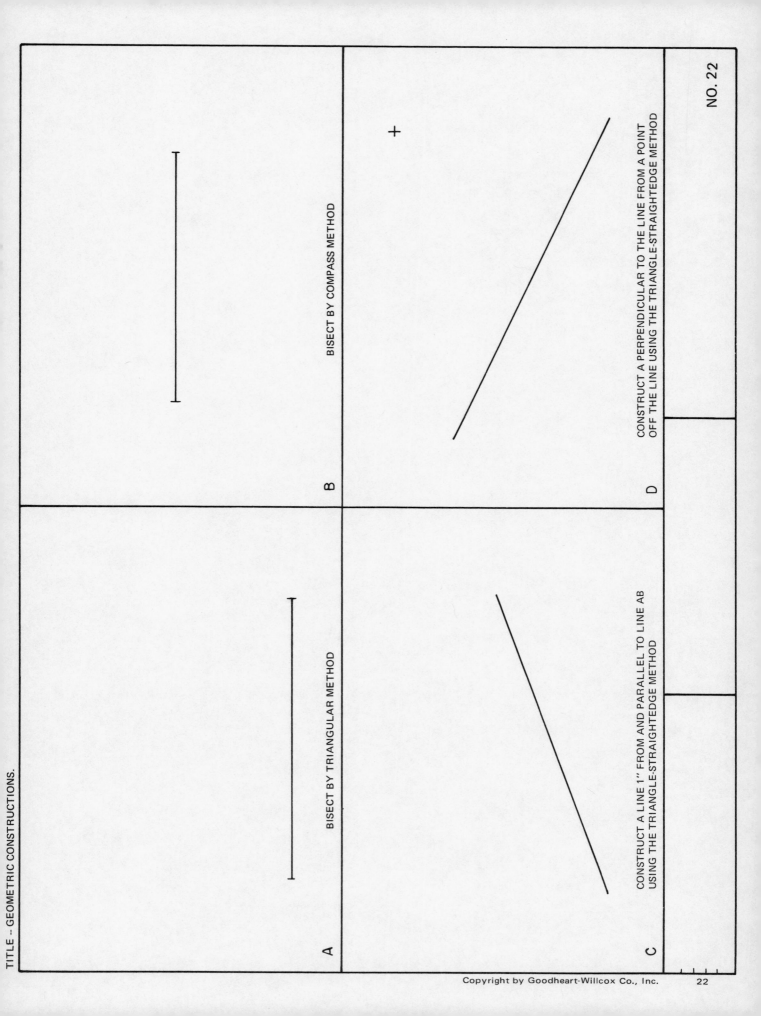

A

BISECT BY TRIANGULAR METHOD

B

BISECT BY COMPASS METHOD

C

CONSTRUCT A LINE 1" FROM AND PARALLEL TO LINE AB
USING THE TRIANGLE-STRAIGHTEDGE METHOD

D

CONSTRUCT A PERPENDICULAR TO THE LINE FROM A POINT
OFF THE LINE USING THE TRIANGLE-STRAIGHTEDGE METHOD

NO. 22

TITLE -- GEOMETRIC CONSTRUCTIONS.

A

B

A

DIVIDE LINE AB GEOMETRICALLY INTO SEVEN EQUAL PARTS USING VERTICAL
LINE METHOD

A

C

B

B

BISECT ANGLE ABC

C

TRANSFER ANGLE ABC FROM ABOVE TO THIS SECTION IN
A POSITION REVOLVED CLOCKWISE OF APPROX 90°

D

A

B

CONSTRUCT A PERPENDICULAR TO LINE AB AT POINT
A USING THE 3-4-5 METHOD.   LET SIDE AB = 4 UNITS

NO. 23

23

A ———————— B

A   GIVEN SIDE AB, ∠A = 37°, ∠B = 70°. CONSTRUCT A TRIANGLE.

B   GIVEN SIDES AB = 3, BC = $1\frac{1}{4}$ AND CA = $2\frac{1}{8}$. CONSTRUCT A TRIANGLE. MEASURE EACH ANGLE.

C   GIVEN SIDES AB = $1\frac{1}{2}$, AC = 2 AND ∠A = 30°. CONSTRUCT A TRIANGLE.

D   GIVEN SIDE AB = $2\frac{3}{4}$. CONSTRUCT AN EQUILATERAL TRIANGLE.

NO. 24

TITLE -- GEOMETRIC CONSTRUCTIONS.

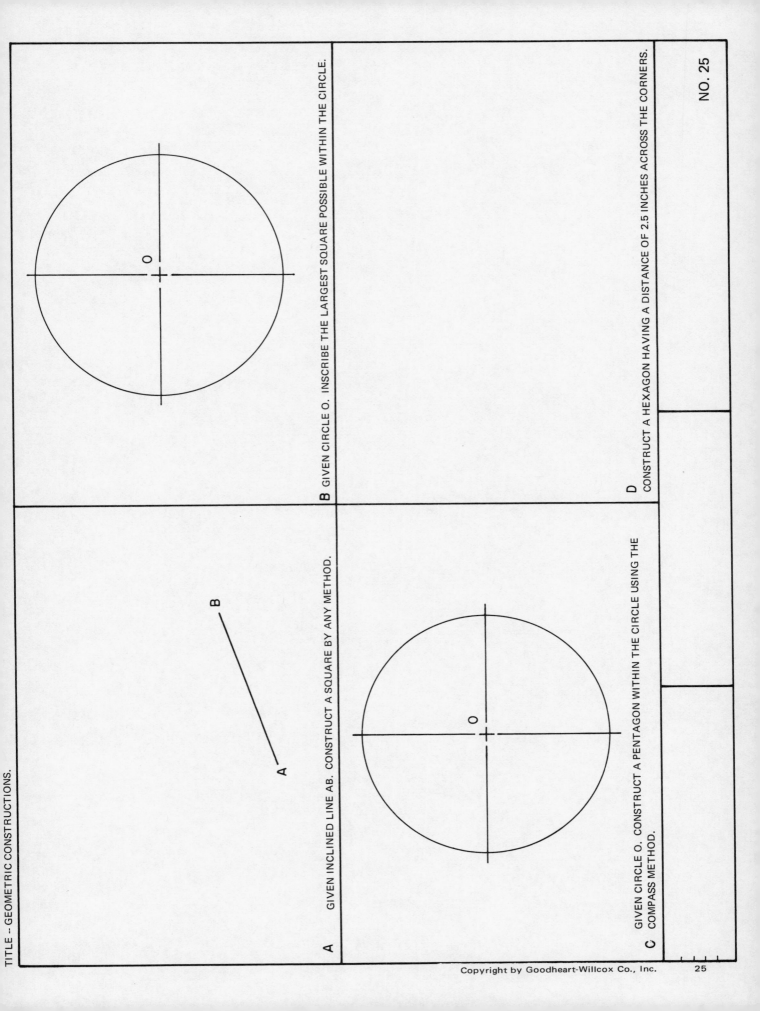

A GIVEN INCLINED LINE AB. CONSTRUCT A SQUARE BY ANY METHOD.

B GIVEN CIRCLE O. INSCRIBE THE LARGEST SQUARE POSSIBLE WITHIN THE CIRCLE.

C GIVEN CIRCLE O. CONSTRUCT A PENTAGON WITHIN THE CIRCLE USING THE COMPASS METHOD.

D CONSTRUCT A HEXAGON HAVING A DISTANCE OF 2.5 INCHES ACROSS THE CORNERS.

NO. 25

A

CONSTRUCT AN OCTAGON WHICH IS $2\frac{1}{2}$" ACROSS THE FLATS.

B

CONSTRUCT A PENTAGON WITH SIDES OF $1\frac{1}{2}$".

C

CONSTRUCT A CIRCLE WHICH PASSES THROUGH THE ABOVE THREE POINTS.

+

+

+

D

LOCATE THE CENTER OF THE ARC.

NO. 26

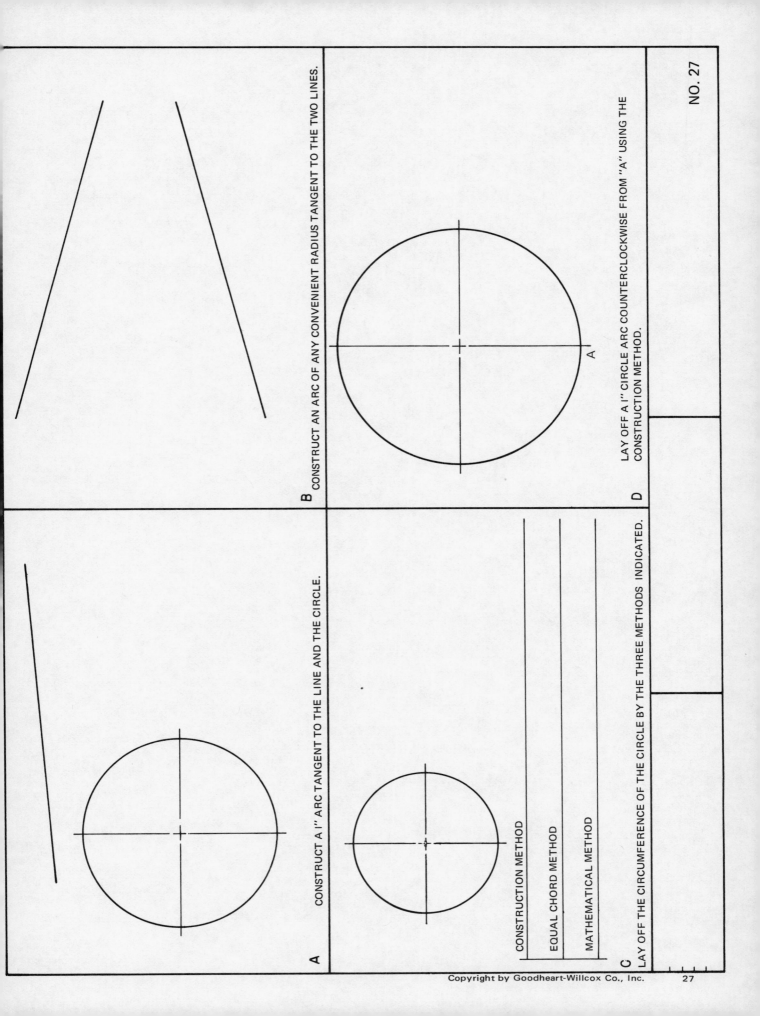

A CONSTRUCT A 1" ARC TANGENT TO THE LINE AND THE CIRCLE.

B CONSTRUCT AN ARC OF ANY CONVENIENT RADIUS TANGENT TO THE TWO LINES.

C LAY OFF THE CIRCUMFERENCE OF THE CIRCLE BY THE THREE METHODS INDICATED.

CONSTRUCTION METHOD

EQUAL CHORD METHOD

MATHEMATICAL METHOD

D LAY OFF A 1" CIRCLE ARC COUNTERCLOCKWISE FROM "A" USING THE CONSTRUCTION METHOD.

A

NO. 27

A DRAW THE OUTLINE OF THE ELLIPTICAL SWIMMING POOL GIVEN THE MAJOR AND MINOR DIAMETERS ABOVE. USE THE FOCI METHOD.

B DRAW AN ELLIPSE GIVEN THE MAJOR AND MINOR DIAMETERS ABOVE. USE THE CONCENTRIC CIRCLE METHOD.

C US:NG AN ELLIPSE TEMPLATE, DRAW AN ELLIPSE AND LETTER THE SIZE AND DEGREE OF THE ELLIPSE.

D SELECT AN APPROPRIATE SCALE AND DRAW A PARABOLIC REFLECTOR FOR A FLOOD LIGHT WHICH HAS A FOCUS OF 5" AND A RISE OF 4".

NO. 28

A CONSTRUCT A HYPERBOLA AND ITS ASYMPTOTES USING THE FOCI METHOD.
GIVEN: HORIZONTAL TRANSVERSE AXIS = .75"; FOCI 1.25" APART.

B CONSTRUCT AN EQUILATERAL HYPERBOLA THRU POINT P.

C DRAW ONE REVOLUTION OF A SPIRAL OF ARCHIMEDES WITH THE GENERATING
POINT MOVING UNIFORMLY IN A COUNTERCLOCKWISE DIRECTION AND AWAY
FROM THE CENTER AT THE RATE OF 1.25" PER REVOLUTION.

D CONSTRUCT THE INVOLUTE OF A POINT STARTING AT THE APEX OF A .50"
EQUILATERAL TRIANGLE FOR ONE REVOLUTION CLOCKWISE.

NO. 29

1

2

3

4

5

6

7

8

NO. 30

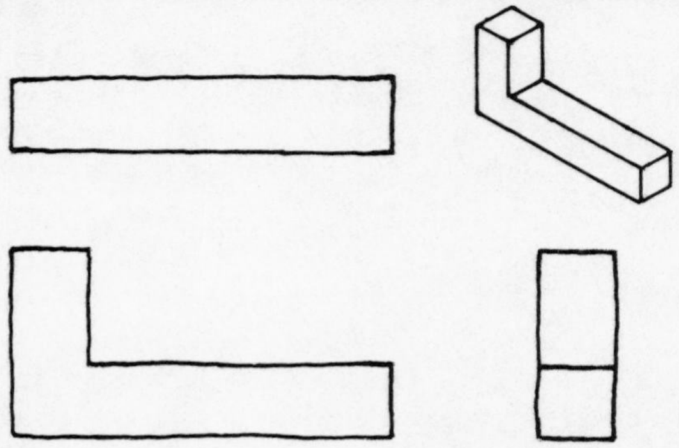

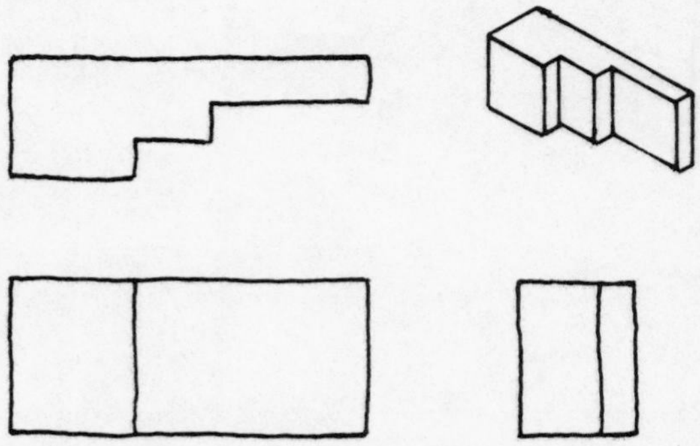

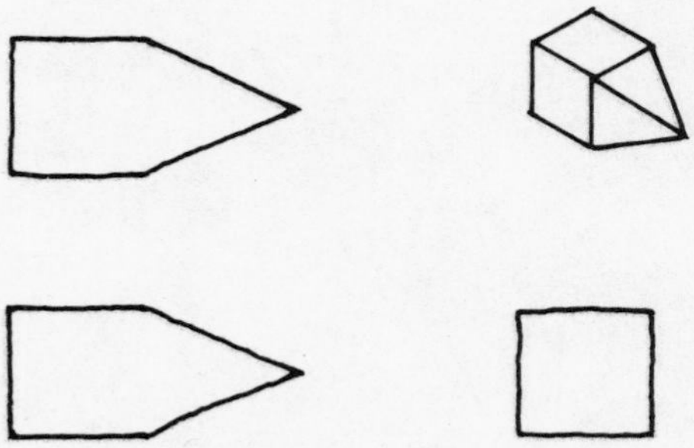

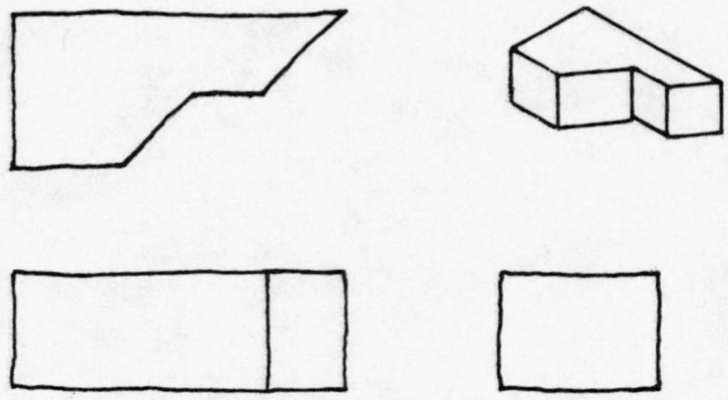

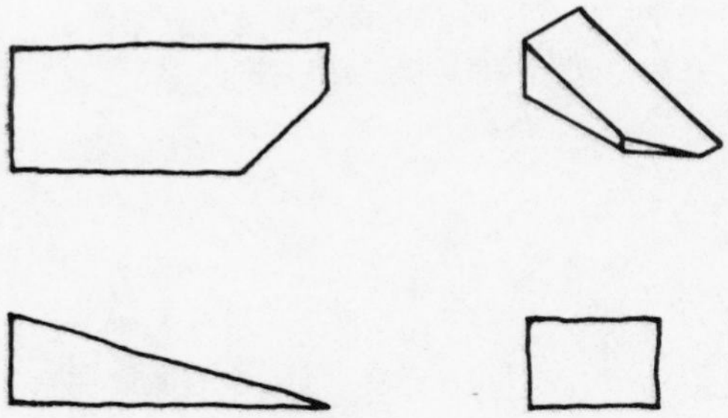

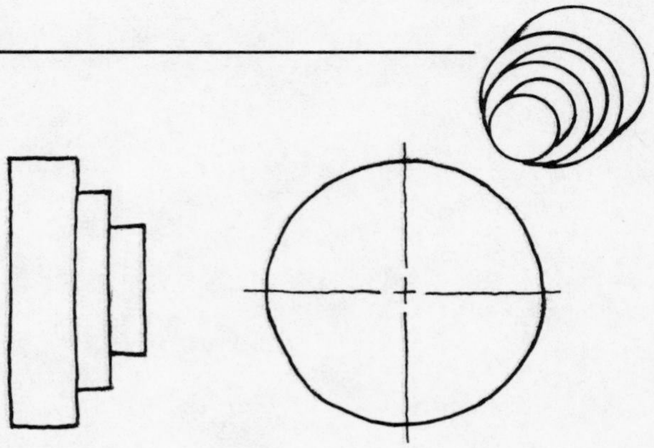

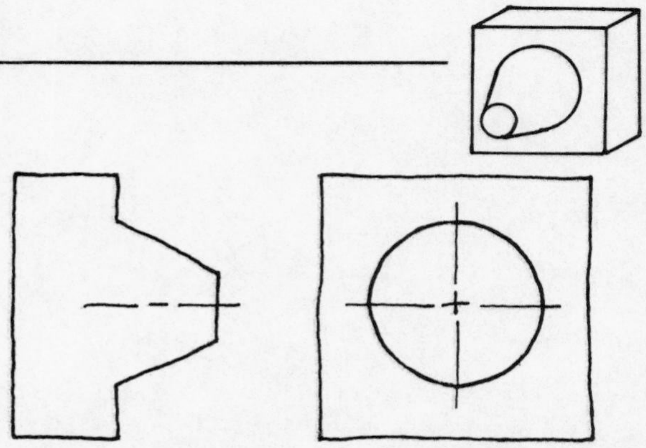

30

1

2

3

4

5

6

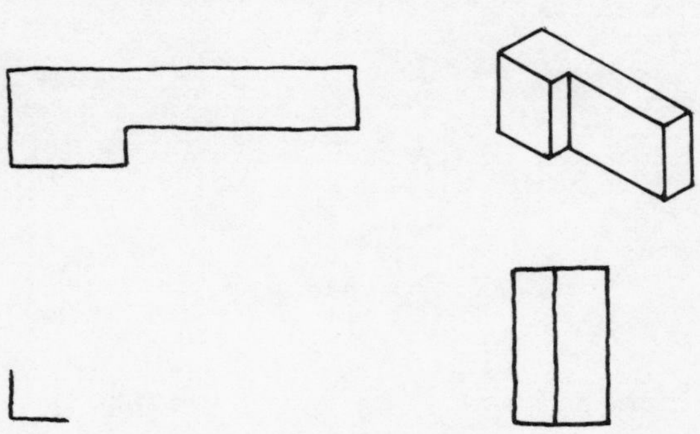

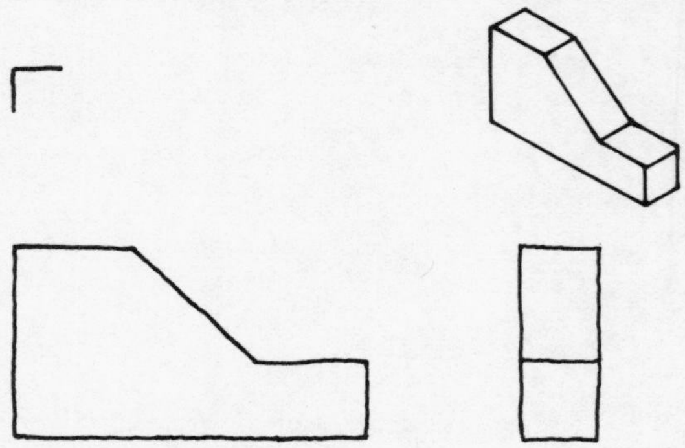

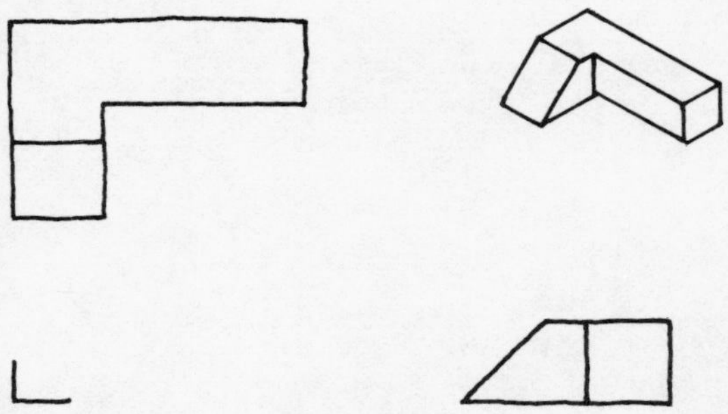

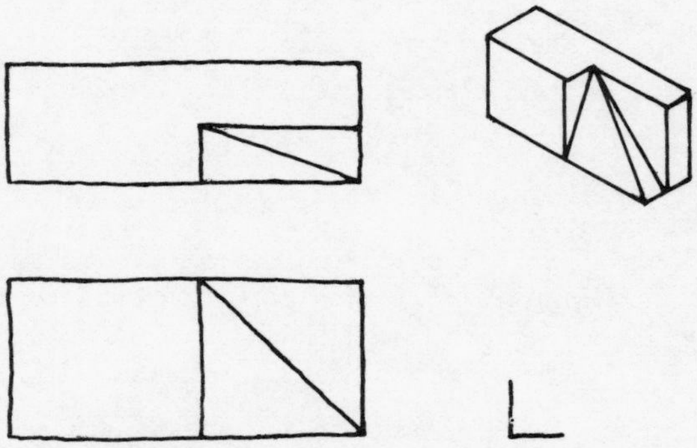

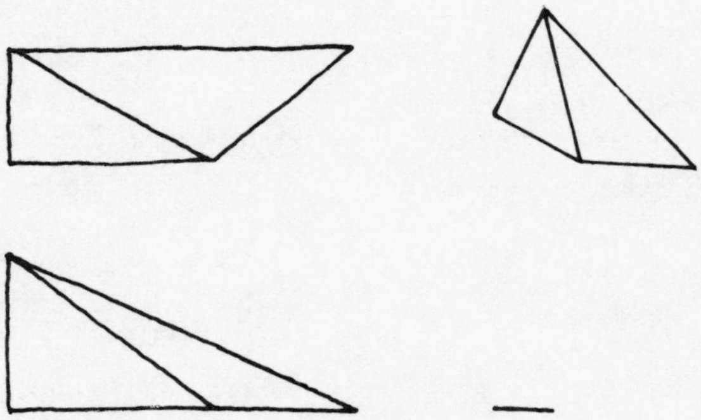

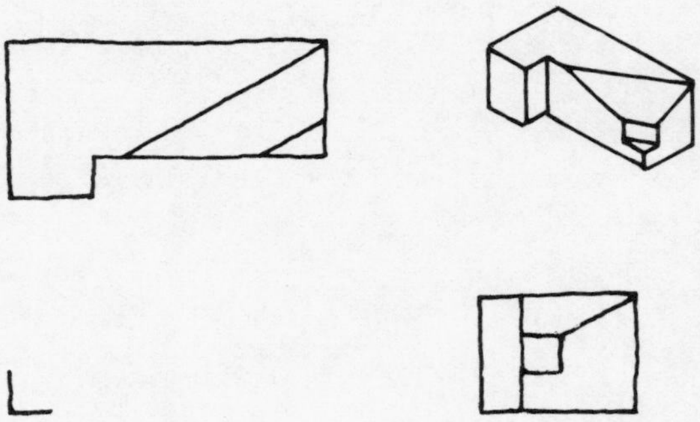

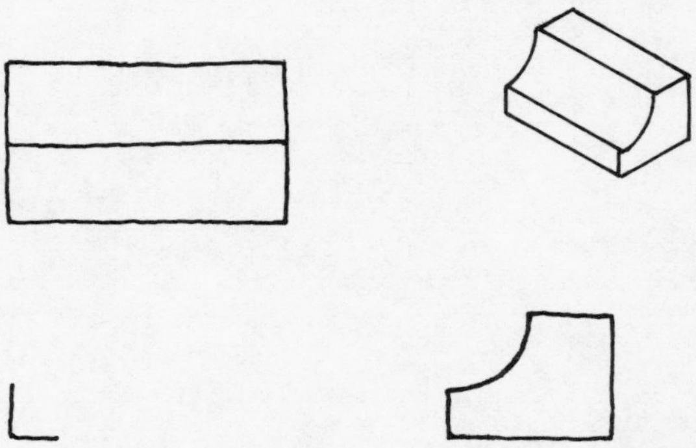

7

8

.281 DRILL, THRU
2 HOLES

.375

3.00

3.00

1.00

.50

1.87 DRILL

1.50

1.88

3.75

END CLAMP

SELECT AN APPROPRIATE SCALE AND MAKE AN INSTRUMENT DRAWING OF THE OBJECT.

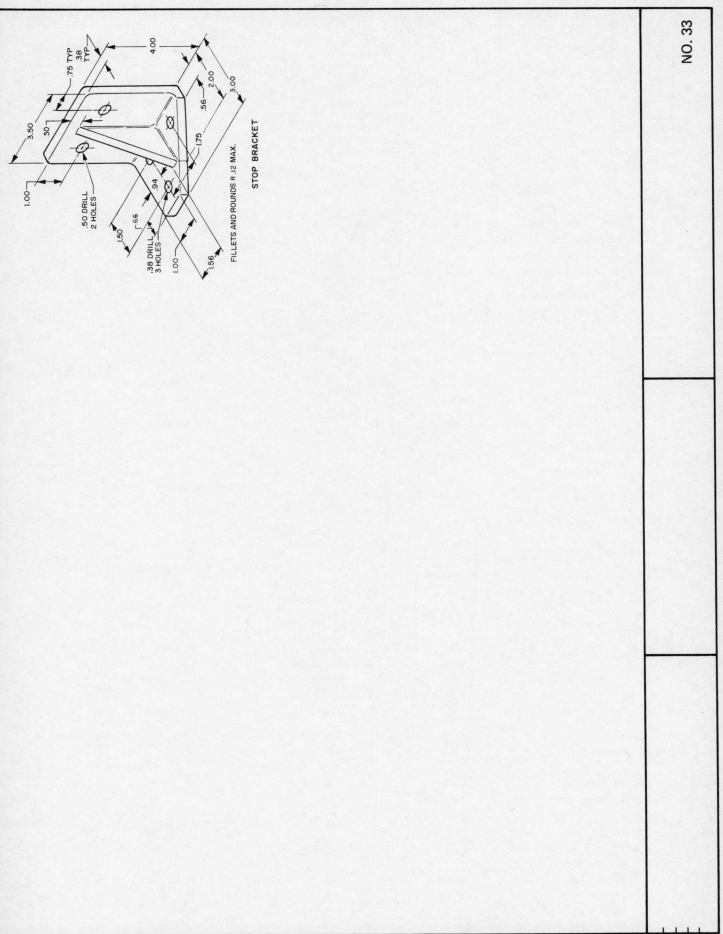

STOP BRACKET

.75 TYP
.38 TYP
4.00
2.00
3.00
3.50
.50
.56
1.75
1.00
.50 DRILL
2 HOLES
.94
.66
.38 DRILL
3 HOLES
1.50
1.00
1.56
FILLETS AND ROUNDS R .12 MAX.

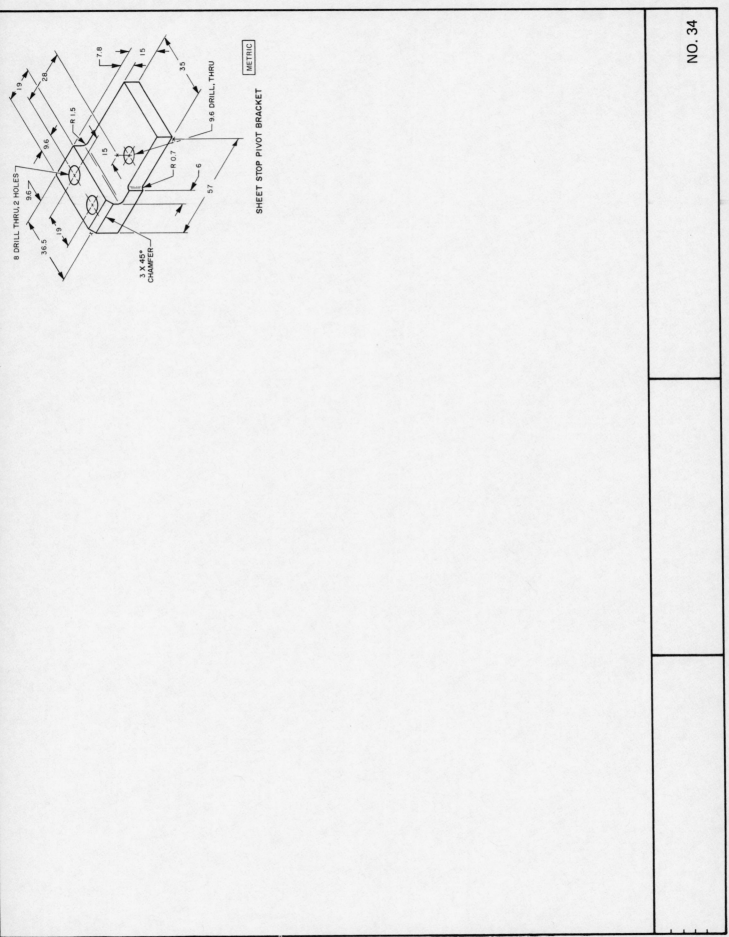

8 DRILL THRU, 2 HOLES

R 1.5

7.8

15

19

28

9.6

9.6

9.6

36.5

19

3 X 45° CHAMFER

R 0.7

6

35

15

9.6 DRILL, THRU

57

SHEET STOP PIVOT BRACKET

METRIC

DO NOT DIMENSION.

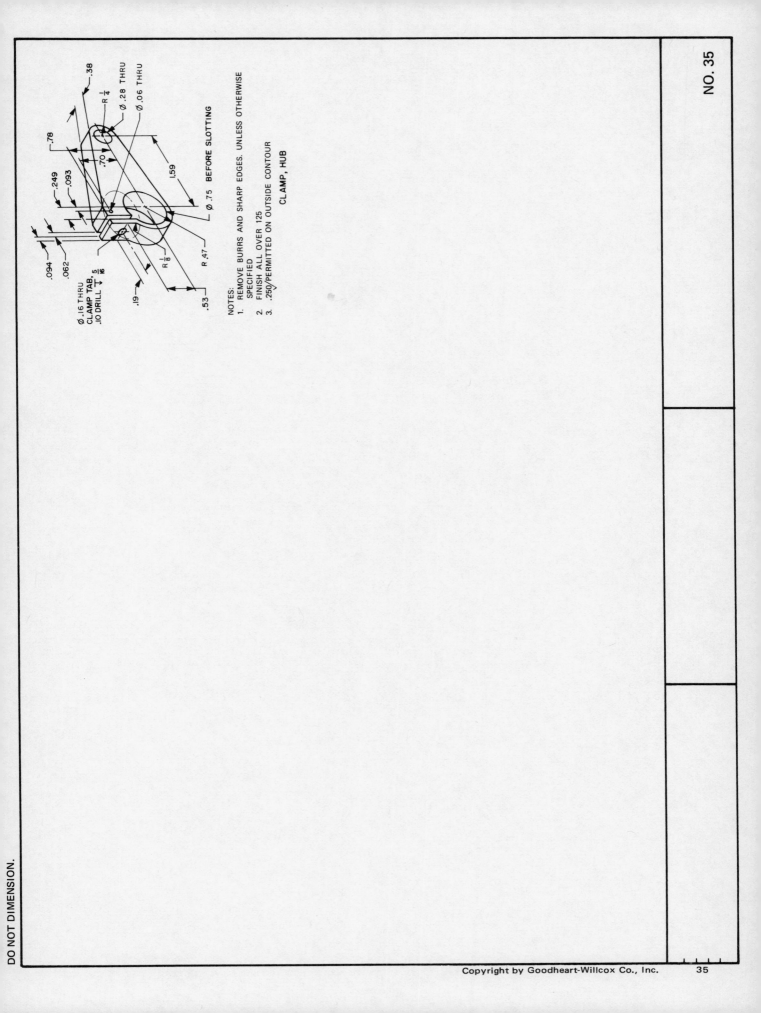

.38

R 1/4

Ø .28 THRU

Ø .06 THRU

.78

.70

1.59

Ø .75   BEFORE SLOTTING

.249

.093

.094

.062

Ø .16 THRU
CLAMP TAB, 5/16
.10 DRILL ↓

.19

R 1/8

R .47

.53

NOTES:
1. REMOVE BURRS AND SHARP EDGES. UNLESS OTHERWISE
   SPECIFIED
2. FINISH ALL OVER 125
3. .250√ PERMITTED ON OUTSIDE CONTOUR

CLAMP, HUB

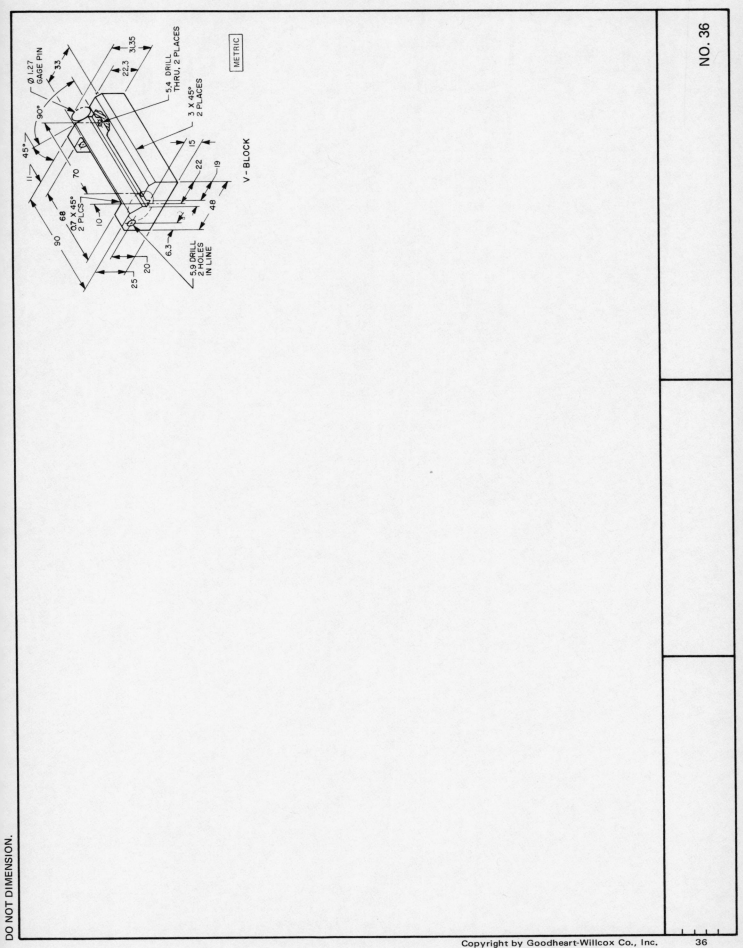

Ø 1.27
GAGE PIN

33

3I.35

22,3

90°

45°

II

90

68

70

0,7 X 45°
2 PLCS

10

25

20

6.3

5.9 DRILL
2 HOLES
IN LINE

48

19

22

15

5,4 DRILL
THRU, 2 PLACES

3 X 45°
2 PLACES

METRIC

V - BLOCK

SELECT AN APPROPRIATE SCALE AND MAKE AN INSTRUMENT DRAWING OF THE NECESSARY VIEWS AND A TWO TIMES SIZE REMOVED VIEW OF THE FEATURE CIRCLED. DO NOT DIMENSION.

Ø 22
BOTH SIDES

22
11

5

15.7

22

11

5

R 1.5
2 PLACES

37

6.3

R

48

6.3

7

25

7

6.3   12.5

25

METRIC

HANGER, SINGLE BEARING

NO. 37

Ø 1.75

1.00

9.50

1.00

2.25

R .06

1.06

6.70

.03 X 45° CHAMFER
3 PLACES

R .25

Ø 2.12

.50

.06

Ø 1.24

Ø .69

Ø .25 DRILL
↧ .69

.36

NOTES:
1. ALL DIA'S TO BE CONCENTRIC WITHIN .0003 FIM
2. HEAT TREAT TO R_c 56-60

LOWER STRAIGHT ANVIL

NO. 38

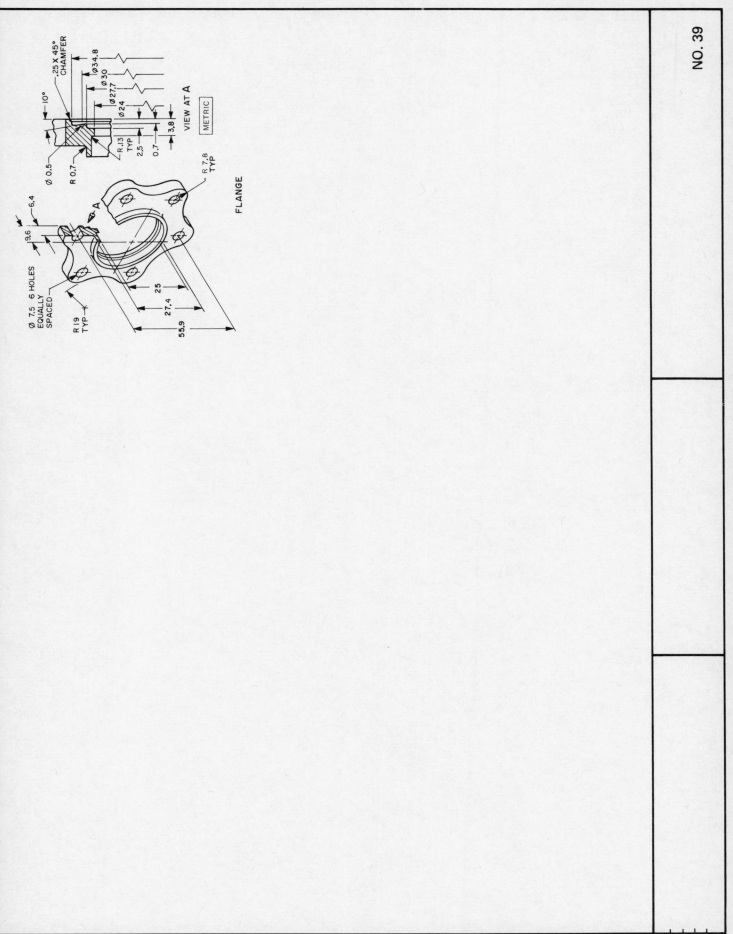

.25 X 45°
CHAMFER

10°

Ø 0.5

R 0.7

Ø34.8

Ø30

Ø27.7

Ø24

R.13
TYP

2.5

0.7

3,8

VIEW AT A

METRIC

R 7.8
TYP

FLANGE

6,4

9,6

Ø 7.5 6 HOLES
EQUALLY
SPACED

A

R 19
TYP

25

27,4

55,9

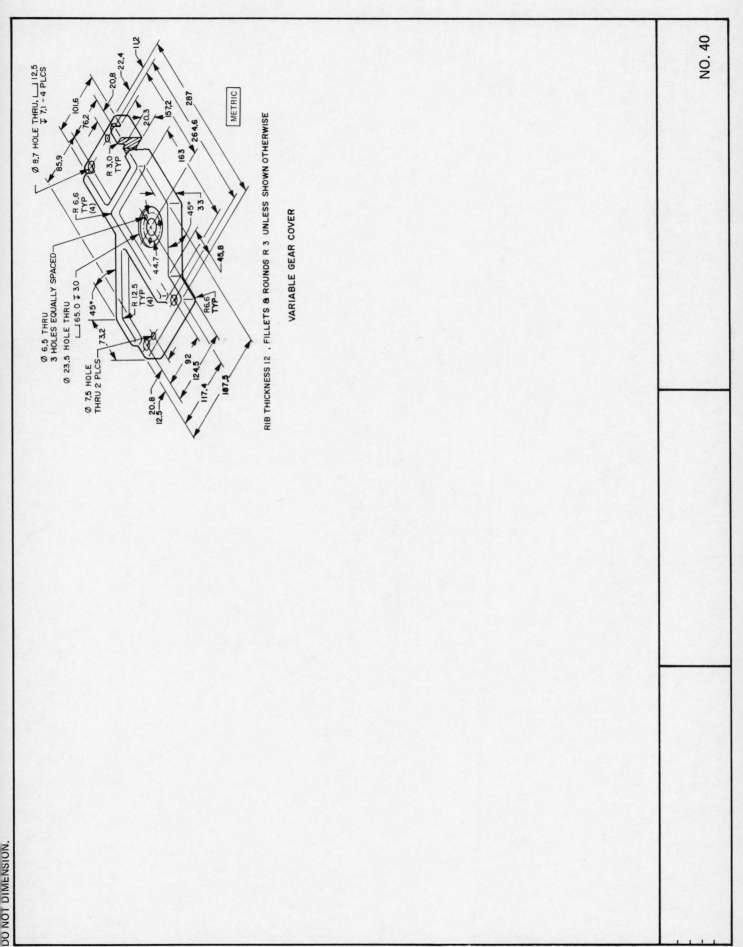

Ø 8.7 HOLE THRU, ⌴ 12.5
⟱ 7.1 - 4 PLCS

112

22.4

20.8

101.6

76.2

R 3.0
TYP

20.3

157.2

287

85.9

R 6.6
TYP
(4)

264.6

163

45°

45°

33

Ø 6.5 THRU
3 HOLES EQUALLY SPACED

44.7

Ø 23.5 HOLE THRU
⌴ 65.0 ⟱ 3.0

45.8

R 12.5
TYP
(4)

45°

73.2

R6.6
TYP

Ø 7.5 HOLE
THRU 2 PLCS

92

20.8

124.5

12.5

117.4

187.5

METRIC

RIB THICKNESS 12 , FILLETS & ROUNDS R 3 UNLESS SHOWN OTHERWISE

VARIABLE GEAR COVER

NO. 40

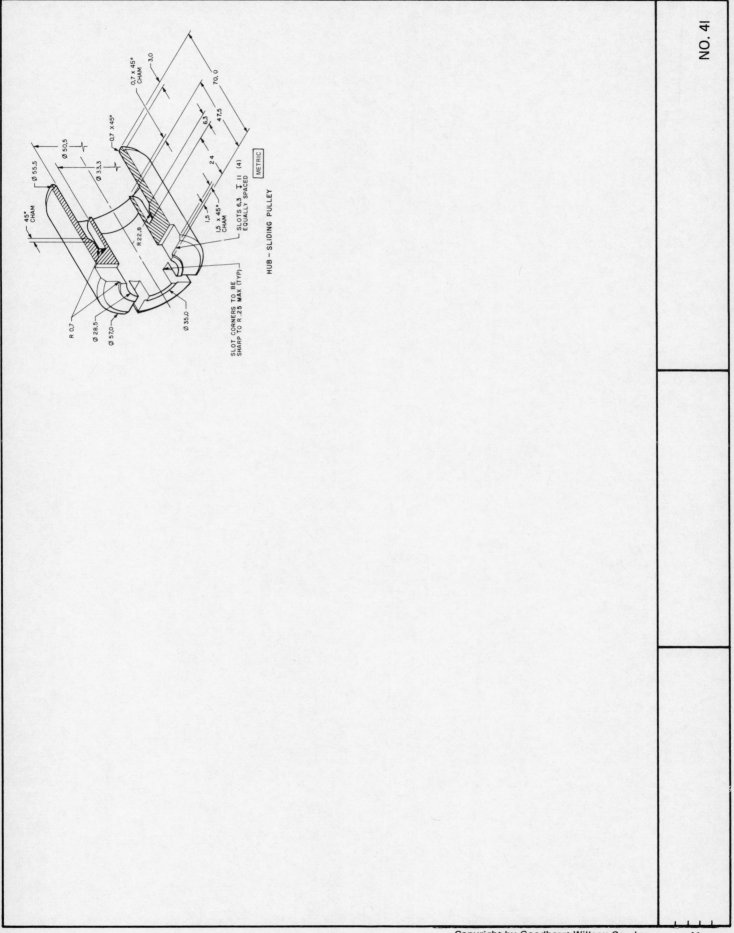

HUB – SLIDING PULLEY

METRIC

45° CHAM

Ø 55.5
Ø 50.5
Ø 33.3

0.7 X 45°

0.7 x 45° CHAM

3.0

70.0

6.3

47.5

2.4

1.5

1.5 x 45° CHAM

SLOTS 6.3 T 11 (4)
EQUALLY SPACED

R 22.8

Ø 35.0

SLOT CORNERS TO BE
SHARP TO R.25 MAX (TYP)

R 0.7

Ø 28.5

Ø 57.0

LOCATE AND SKETCH THE REQUIRED EXTENSION AND DIMENSION LINES AND LEADERS TO DIMENSION THE PARTS SHOWN.
NO DIMENSION FIGURES ARE TO BE INCLUDED.

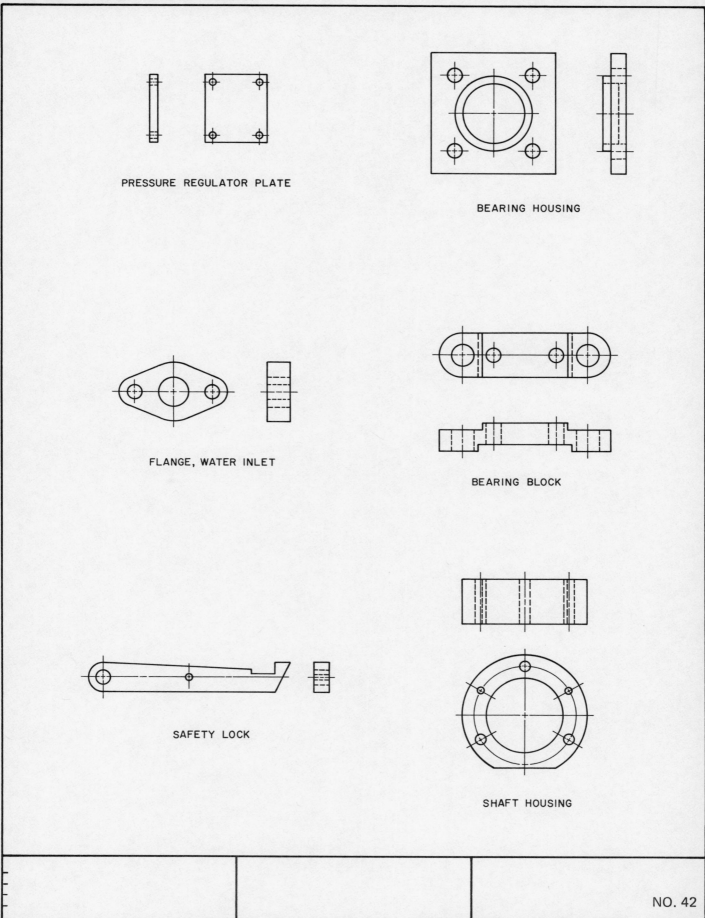

PRESSURE REGULATOR PLATE

BEARING HOUSING

FLANGE, WATER INLET

BEARING BLOCK

SAFETY LOCK

SHAFT HOUSING

NO. 42

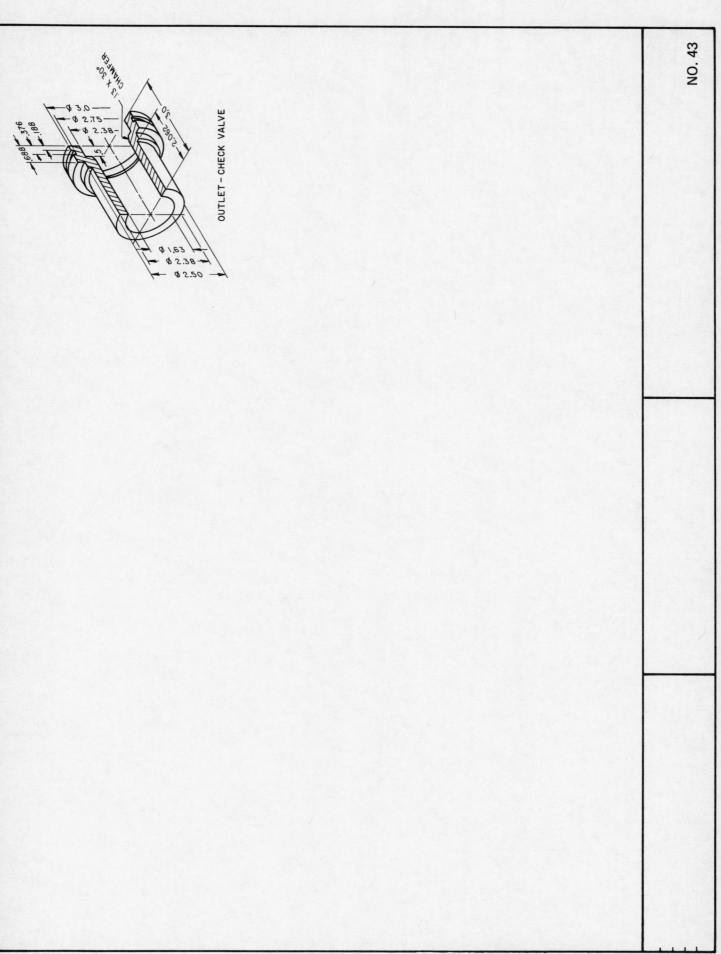

OUTLET – CHECK VALVE

⌀ 3.0
⌀ 2.75
⌀ 2.38
⌀ 1.83
⌀ 2.38
⌀ 2.50

2.062
3.0

13 X 30°
CHAMFER

.688
.376
.188

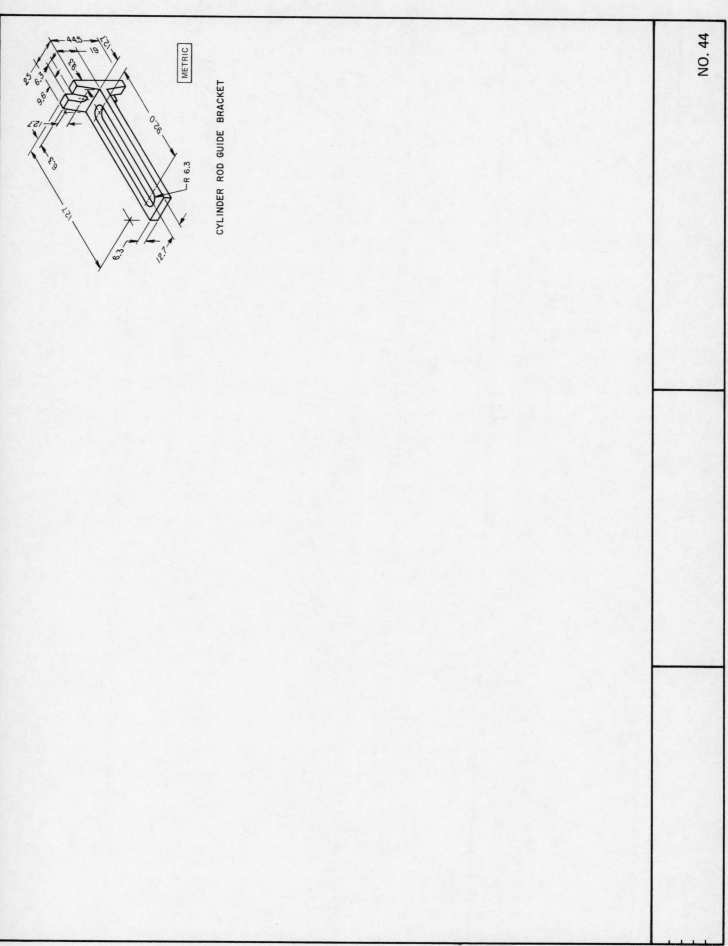

CYLINDER ROD GUIDE BRACKET

METRIC

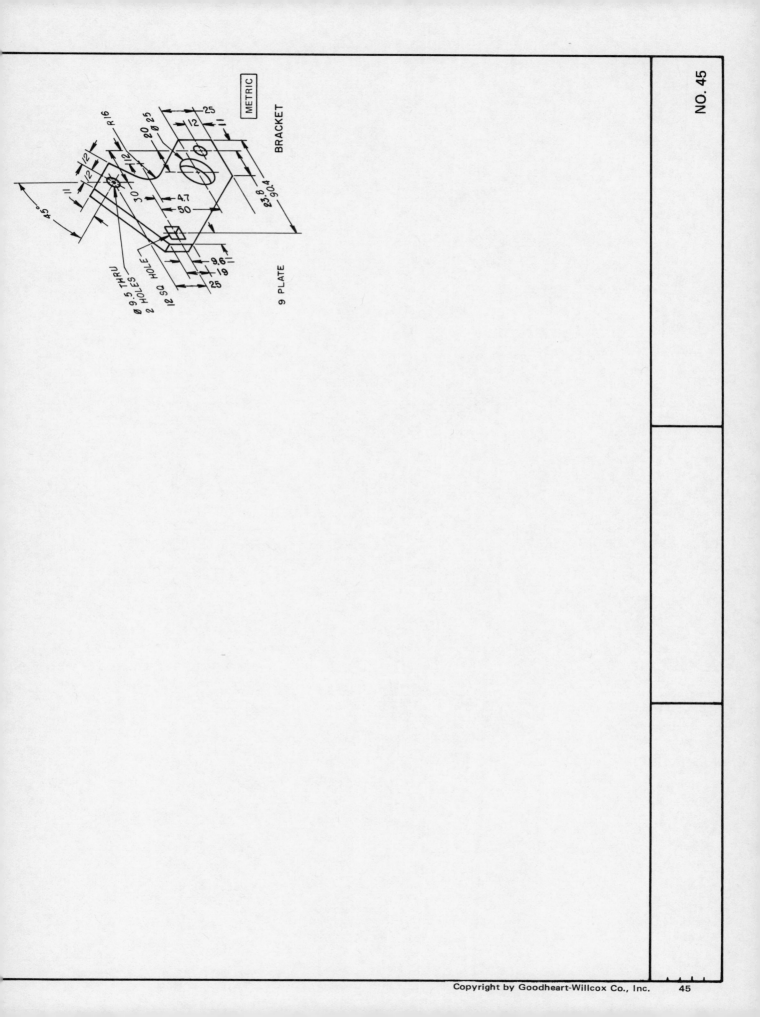

METRIC

BRACKET

9 PLATE

R 16

Ø 25

45°

Ø 9.5 THRU
2 HOLES

12 SQ HOLE

25
12
11

20

12
12
11

30
4.7
50

23.8
90.4

9.6
19
25

11

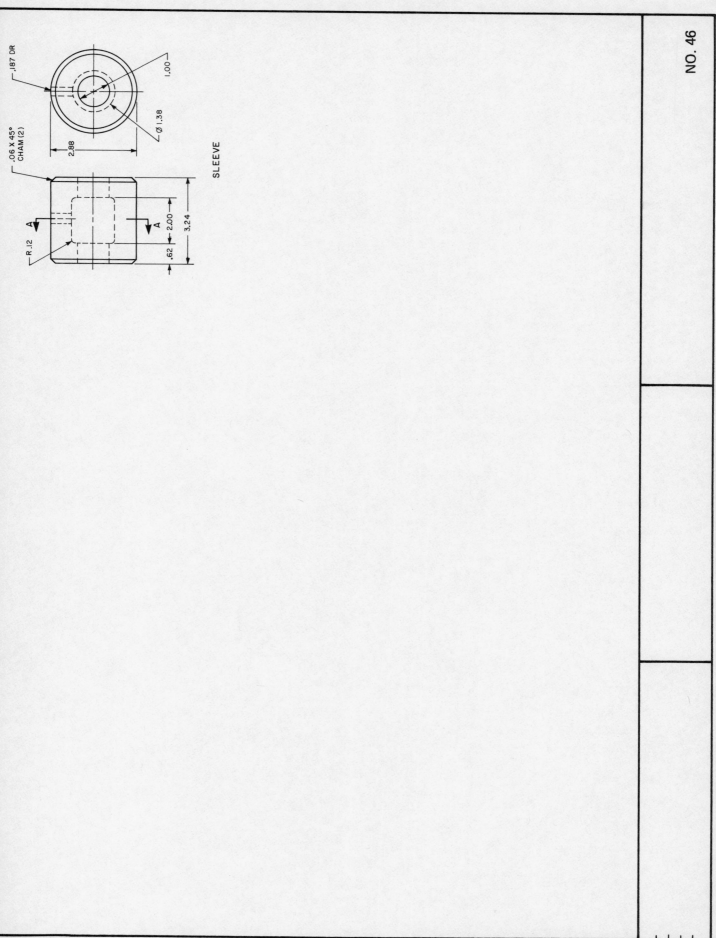

.187 DR

1.00

⌀ 1.38

.06 X 45°
CHAM (2)

2.88

SLEEVE

R .12

A

A

3.24

2.00

.62

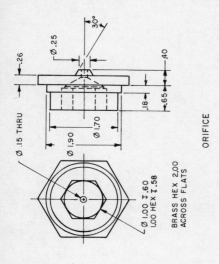

ORIFICE

Ø .25

.26

30°

.40

.65

.18

Ø 1.70

Ø 1.90

Ø .15 THRU

Ø 1.00 ⌴ .60
1.00 HEX ⌴ .58

BRASS HEX 2.00
ACROSS FLATS

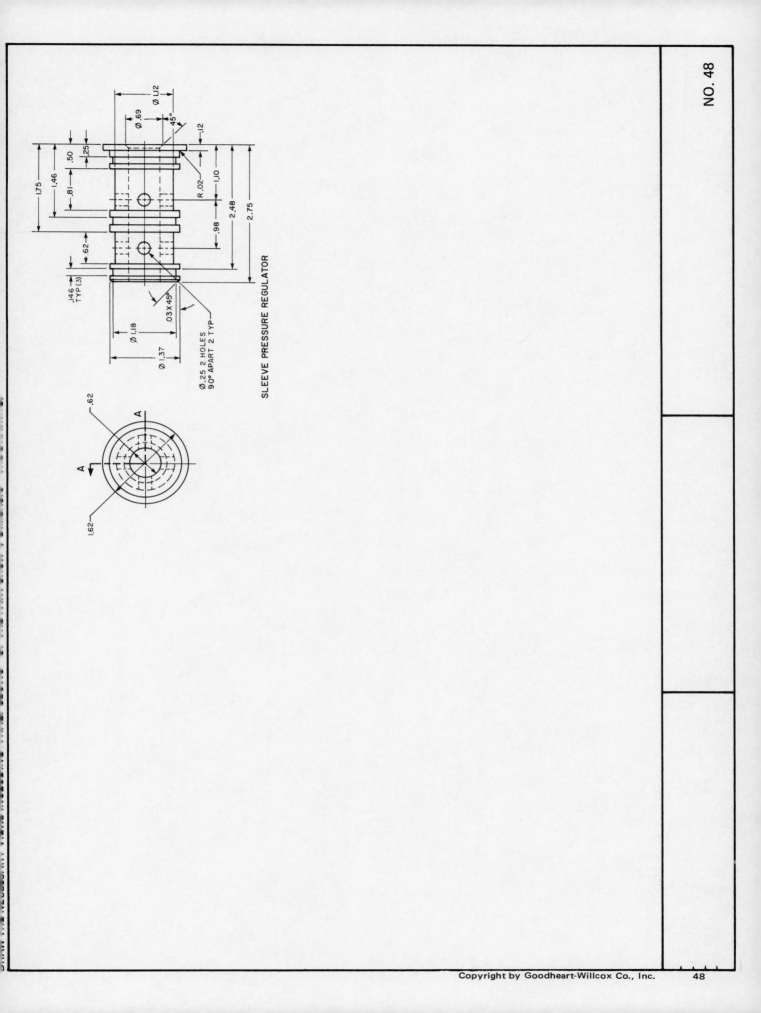

SLEEVE PRESSURE REGULATOR

Ø 1.12
Ø .69
45°
.12
.50
.25
1.75
1.46
.81
R .02
1.10
2.48
2.75
.98
.62
.146 TYP (3)
Ø 1.18
.03 X 45°
Ø 1.37
Ø .25 2 HOLES 90° APART 2 TYP.

A
.62
1.62
A

PISTON

DETAIL A

SEE DETAIL A

METRIC

21.3

Ø 66.5

Ø 59.5

Ø25.0

0.7

1.5

4.1

13.5

17.5

A

A

DRAW THE NECESSARY VIEWS INCLUDING REVOLVED SECTIONS OF THE APPROPRIATE FEATURES. DIMENSION THE DRAWING.

R 3
R 38
15°
5°
25
∅ 50
∅ 38
∅ 14.287
12.5
SECTION B-B

∅ 30
∅ 24
9
24
STEEL INSERT
2-4646-101
CAST IN PLACE
(REF BELOW)

63.5
26.9
R 4.5
27
R 101.6

20.5
9.4
R 4.5
6.4
∅ 22.23
R 4
12.5

R 3
12.5
SECTION C-C
150.6
R 203
R 152
44.5
137.9

A

METRIC

HANDWHEEL

R 4
R 3
C
22
A
17.3
B C
16.5
B
4.7
KWY
152
∅ 49 TO
CLEAN UP

DRAW THE NECESSARY VIEWS INCLUDING A REMOVED SECTIONAL VIEW OF THE FEATURES INDICATED. USE SECTIONAL VIEWS FOR REGULAR VIEWS WHERE CLARITY WILL BE IMPROVED.

Ø 2.81
Ø 2.25
Ø 1.31
R .12
.25
2.00
1.03
.94
A
R .02 AT BASE OF GROOVE
Ø 2.9
Ø 2.31

LOWER ECCENTRIC

.23
1.88 BC
.03 X 45°
CHAMFER
BOTH ENDS
2 HOLES .312 DR
45°

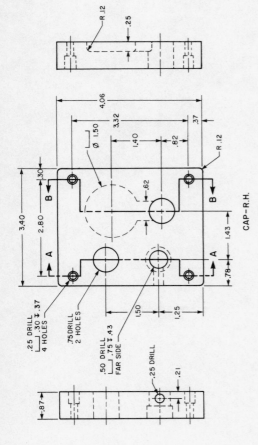

R .12
.25

4.06
3.32
.37

Ø 1.50
1.40
.82
R .12

.30
B
B

.62

3.40
2.80

A
A

1.43

.78

.25 DRILL
⌴ .30 ⊼ .37
4 HOLES

.75 DRILL
2 HOLES

.50 DRILL
.75 ⊼ 1.43
FAR SIDE

.25 DRILL

1.50
1.25

CAP–R.H.

.87

.21

COVER, PITOT OVERRIDE

2.49
1.12
1.10
1.62
1.41
1.12
1.03
2.51
1.75
A
A
CAST RADII .12 UNLESS OTHERWISE SPECIFIED
R .38 5 PLACES
.34 DRILL ⌴ .687 4 PLACES
.31 DRILL THRU
.720
1.75

.30
1.25
Ø .78
Ø 1.25
R .06
R .03
1.28
.94
.69
2.25
.37 DRILL
.90
.06
30°
Ø .375
Ø 1.00

DRAW THE NECESSARY VIEWS INCLUDING A BROKEN-OUT SECTION.

.186
R .015
.016x45°
CHAMFER
R .005 MAX
Ø .093
Ø .073
Ø .052
↧ .044
.060
.230
.325

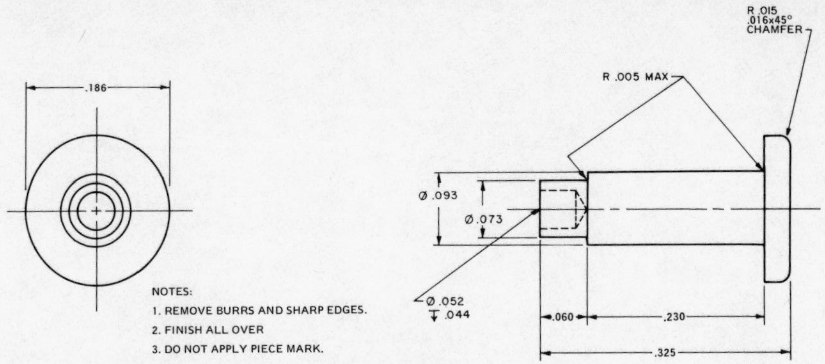

NOTES:
1. REMOVE BURRS AND SHARP EDGES.
2. FINISH ALL OVER
3. DO NOT APPLY PIECE MARK.

PIN, PIVOT

NO. 54

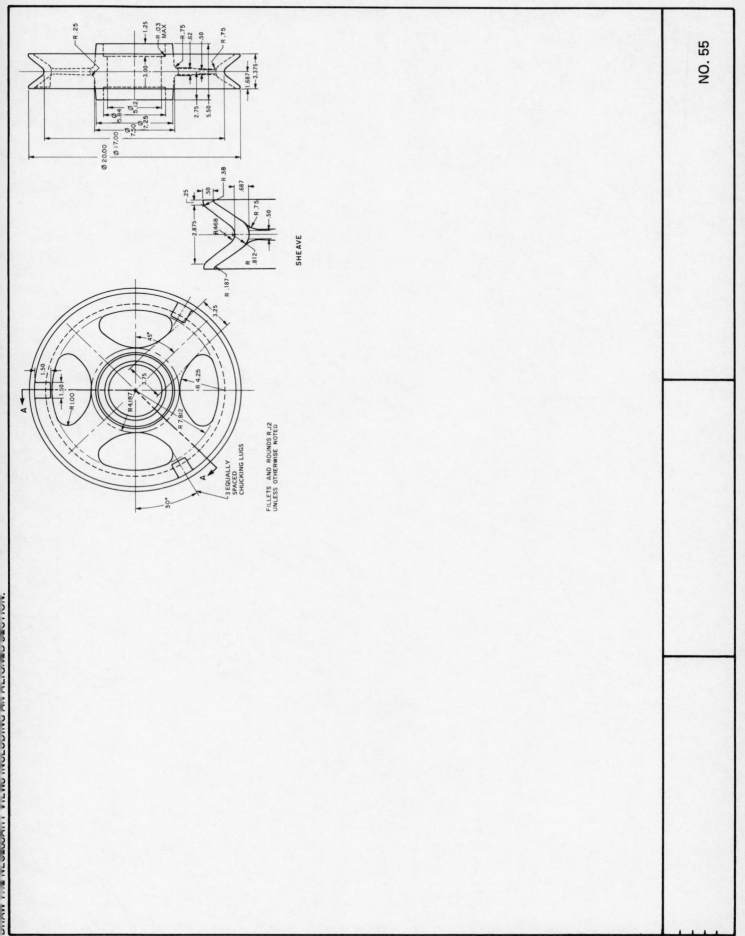

SHEAVE

3 EQUALLY
SPACED
CHUCKING LUGS

FILLETS AND ROUNDS R.I2
UNLESS OTHERWISE NOTED

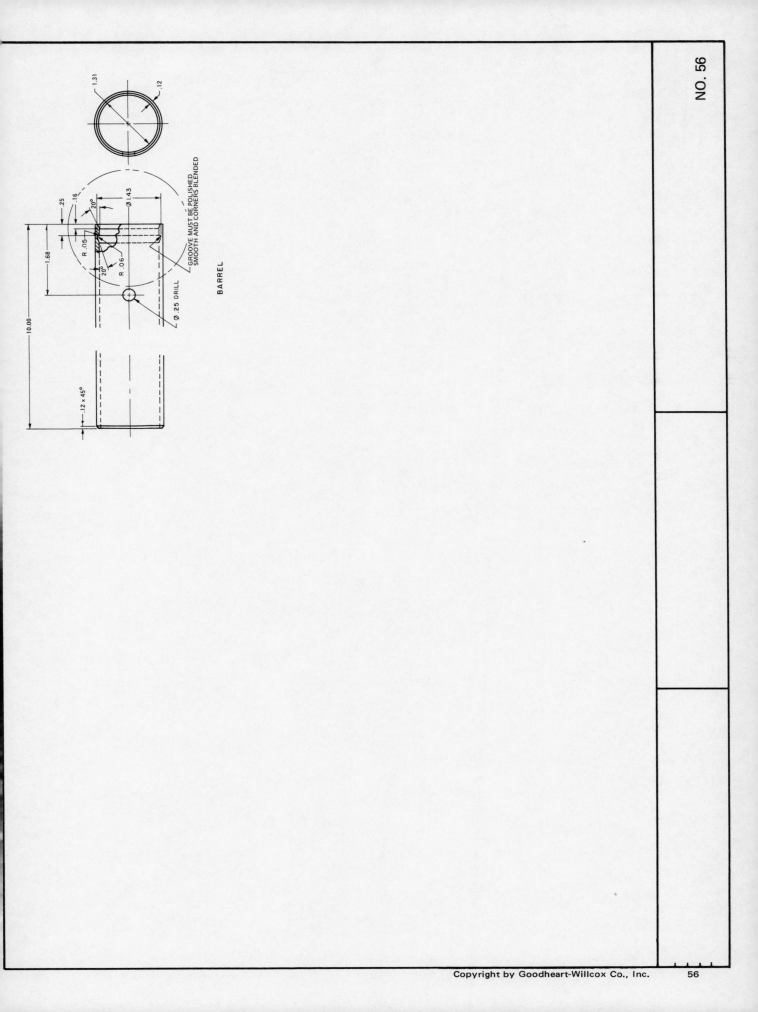

1.31

.12

.25

.16

20°

Ø1.43

R .05

20°

R .06

1.68

10.00

GROOVE MUST BE POLISHED,
SMOOTH AND CORNERS BLENDED

Ø .25 DRILL

BARREL

.12 x 45°

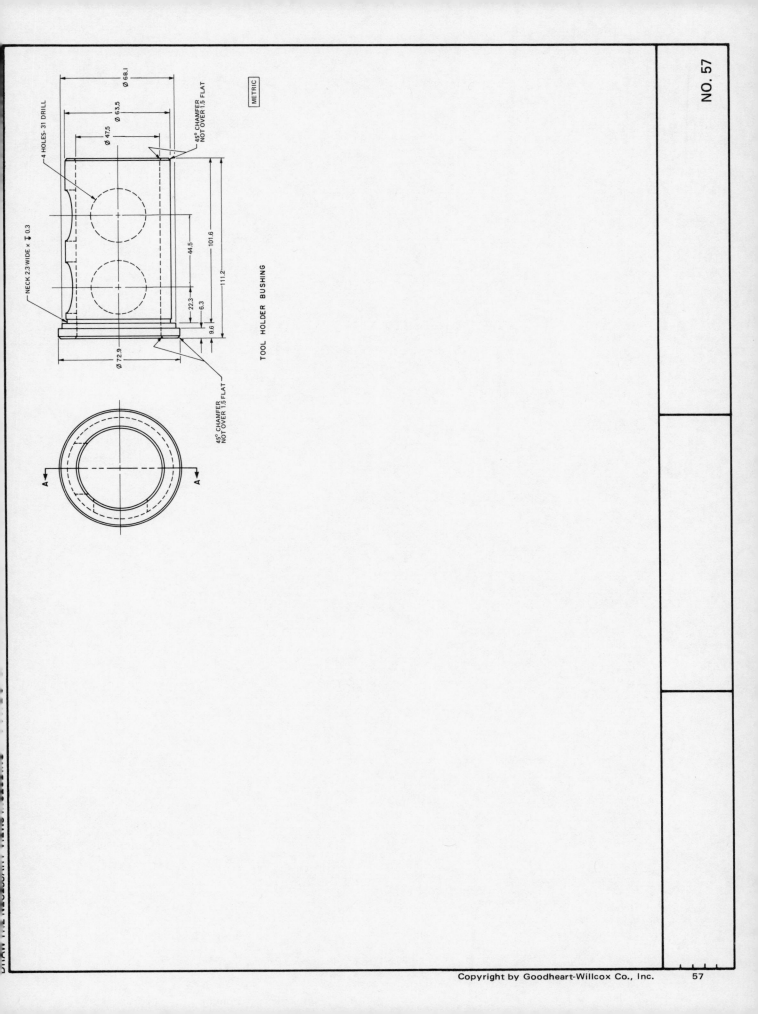

TOOL HOLDER BUSHING

METRIC

Ø 68.1

Ø 63.5

Ø 47.5

4 HOLES- 31 DRILL

45° CHAMFER
NOT OVER 1.5 FLAT

NECK 2.3 WIDE × ↧ 0.3

101.6

111.2

44.5

22.3

6.3

9.6

Ø 72.9

45° CHAMFER
NOT OVER 1.5 FLAT

A

A

DRAW THE NECESSARY VIEWS INCLUDING A FULL SECTION. EITHER METHOD OF SHOWING WEBS IN SECTION MAY BE USED.

Ø 28.5 AREA TO BE
CAST SMOOTH AND FLAT
ONE PLACE

13 DRILL THRU
ONE HOLE

R 83

R 17

R 7.5

R 16

AREA WITHIN PHANTOM
OUTLINE TO BE CAST
SMOOTH AND FLAT
2 PLACES

R 21

R 114

R 103

R 6,3

31.7

18°

45°

11

67.3

255.5

12

6

R 50

68.3

68.3

95.4

47.7

A

A

BRACKET , FAN

METRIC

R 2
57.2
2 PLCS

R 6

R 22

77

73.6

45°

Ø 63.5

Ø 49.3

Ø 46

R 3 TYP

16

20

92.7

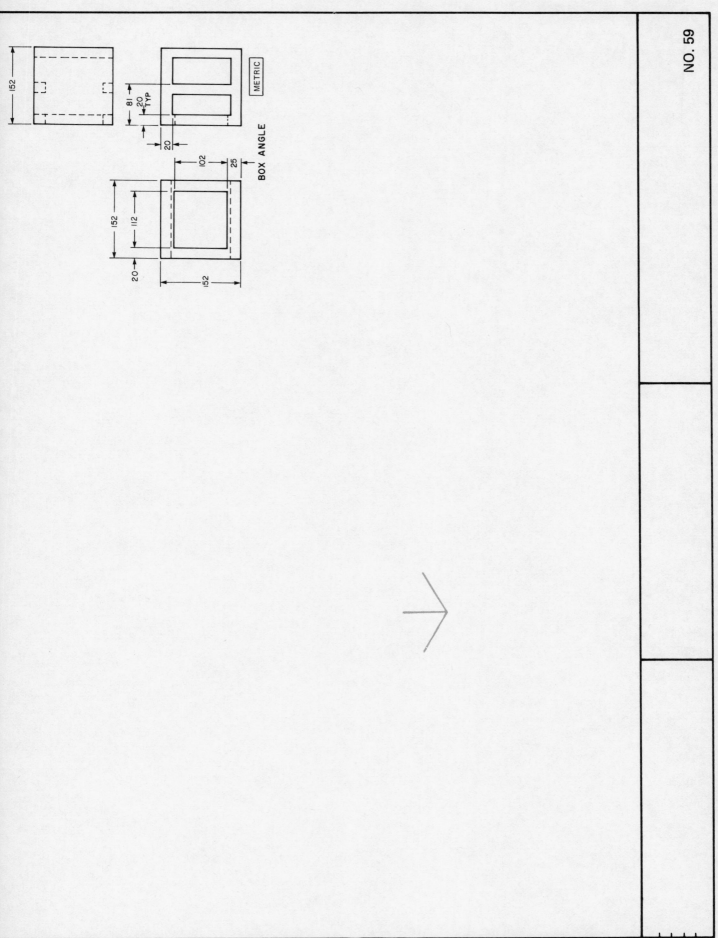

BOX ANGLE

METRIC

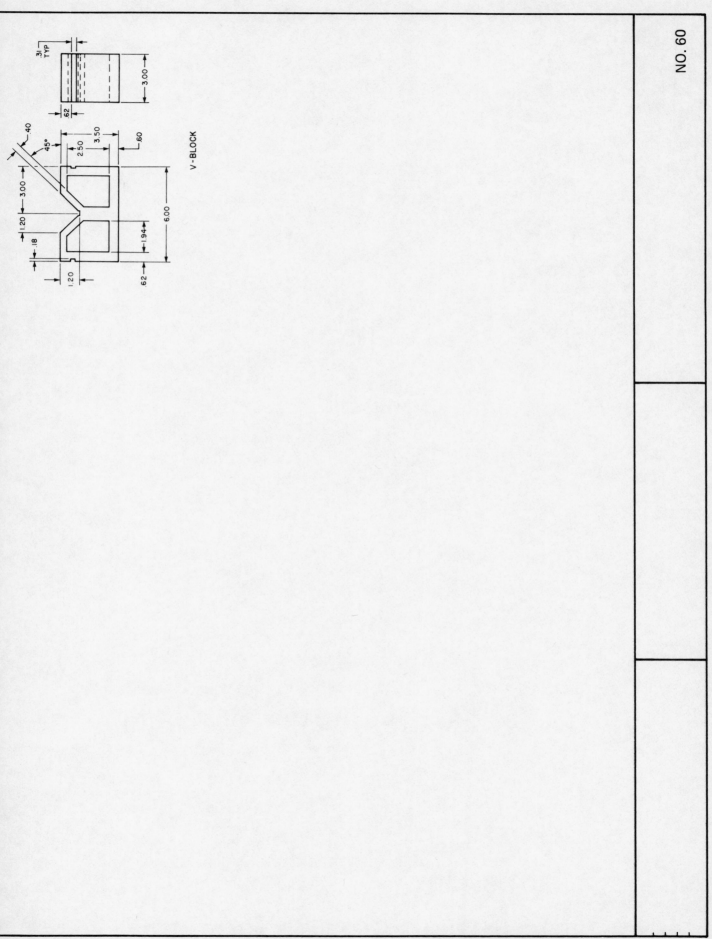

V-BLOCK

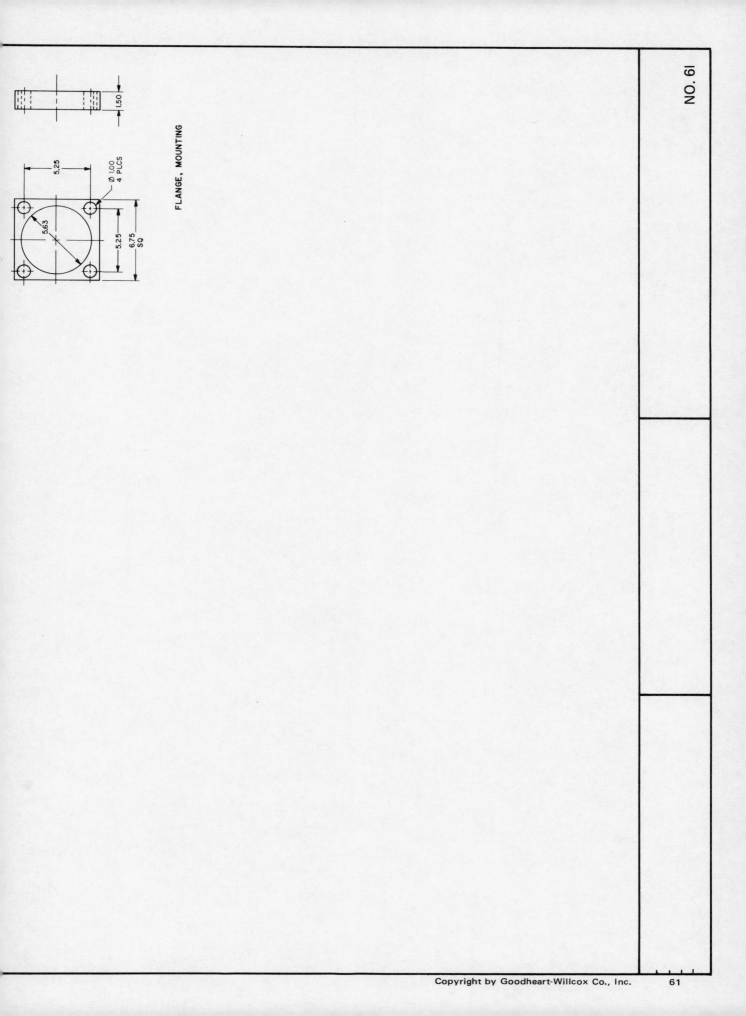

FLANGE, MOUNTING

5.25

Ø 1.00
4 PLCS

5.63

5.25

6.75
SQ

1.50

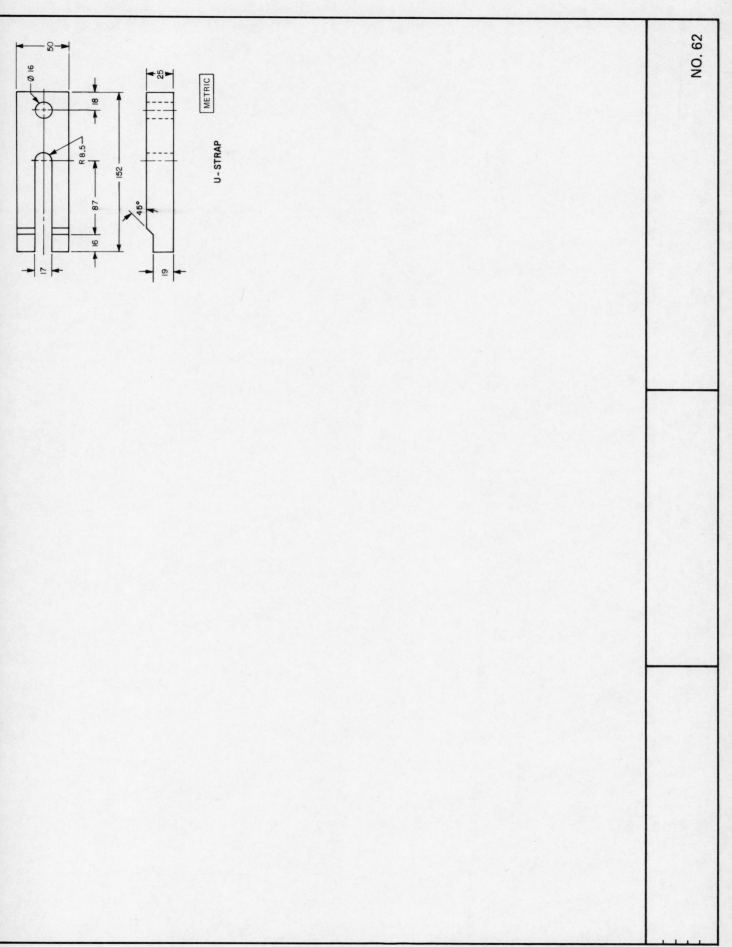

U - STRAP

METRIC

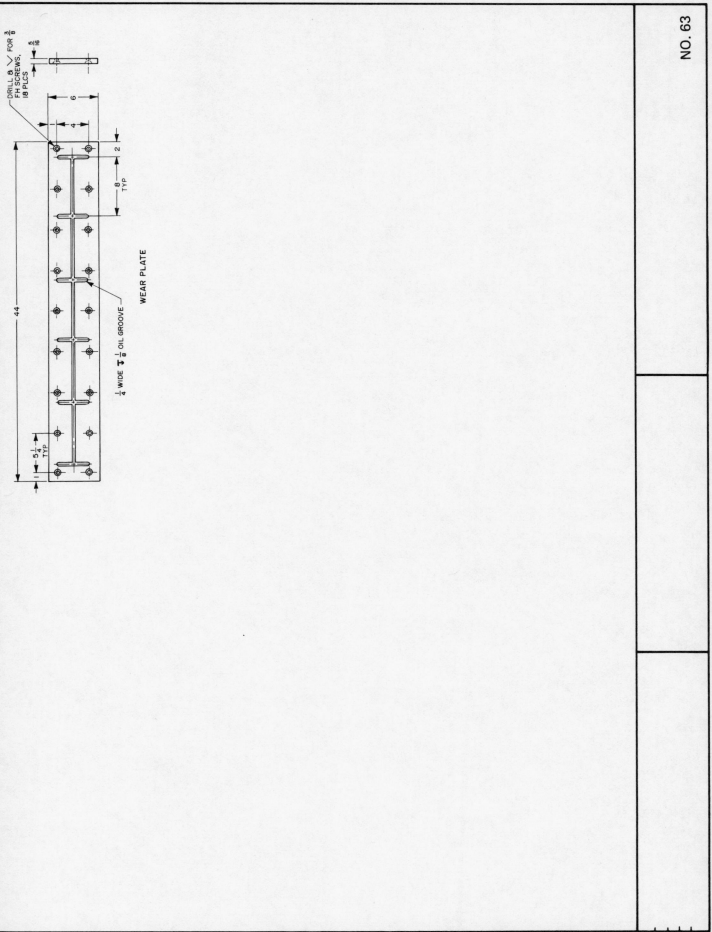

DRILL & ∨ FOR ⅜
FH SCREWS,
18 PLCS

5/16

6

4

2

8
TYP

44

WEAR PLATE

¼ WIDE ⊤ ⅛ OIL GROOVE

5¼
TYP

1

CONSTRUCT A ONE-POINT PERSPECTIVE DRAWING OF THE PART. DO NOT DIMENSION.

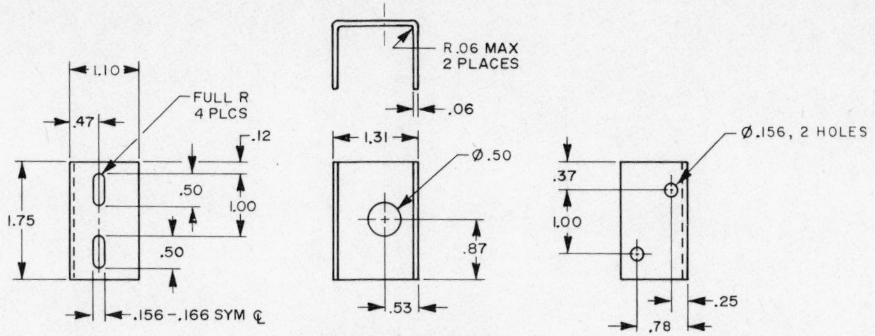

Ø.156, 2 HOLES

.25

.78

.37

1.00

R.06 MAX
2 PLACES

.06

Ø.50

.87

1.31

.53

MOUNTING BRACKET

FULL R
4 PLCS

.12

1.00

.50

.50

.156–.166 SYM ₵

1.10

.47

1.75

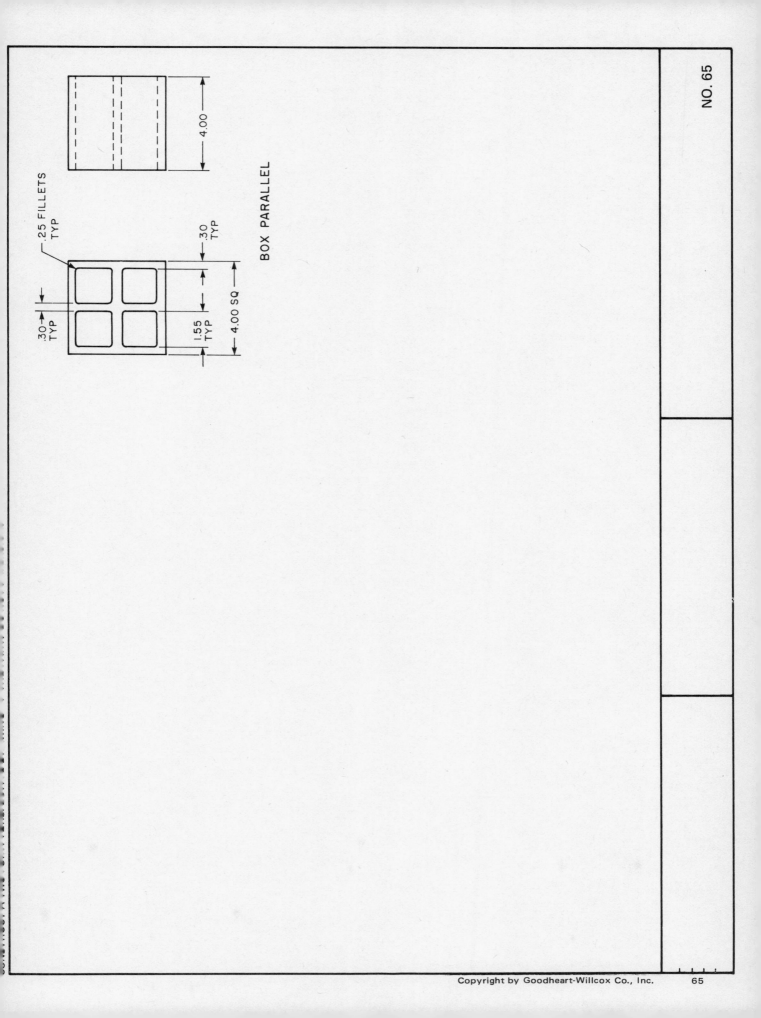

.25 FILLETS
TYP

.30
TYP

.30
TYP

1.55
TYP

4.00 SQ

4.00

BOX PARALLEL

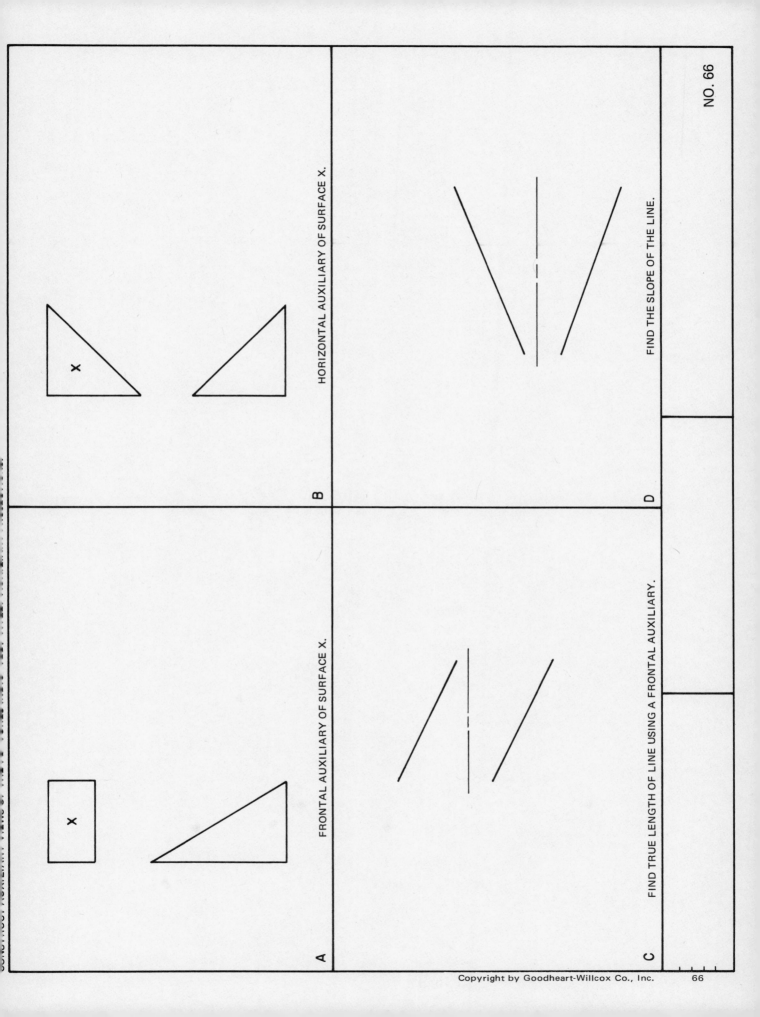

A

FRONTAL AUXILIARY OF SURFACE X.

B

HORIZONTAL AUXILIARY OF SURFACE X.

C

FIND TRUE LENGTH OF LINE USING A FRONTAL AUXILIARY.

D

FIND THE SLOPE OF THE LINE.

NO. 66

66

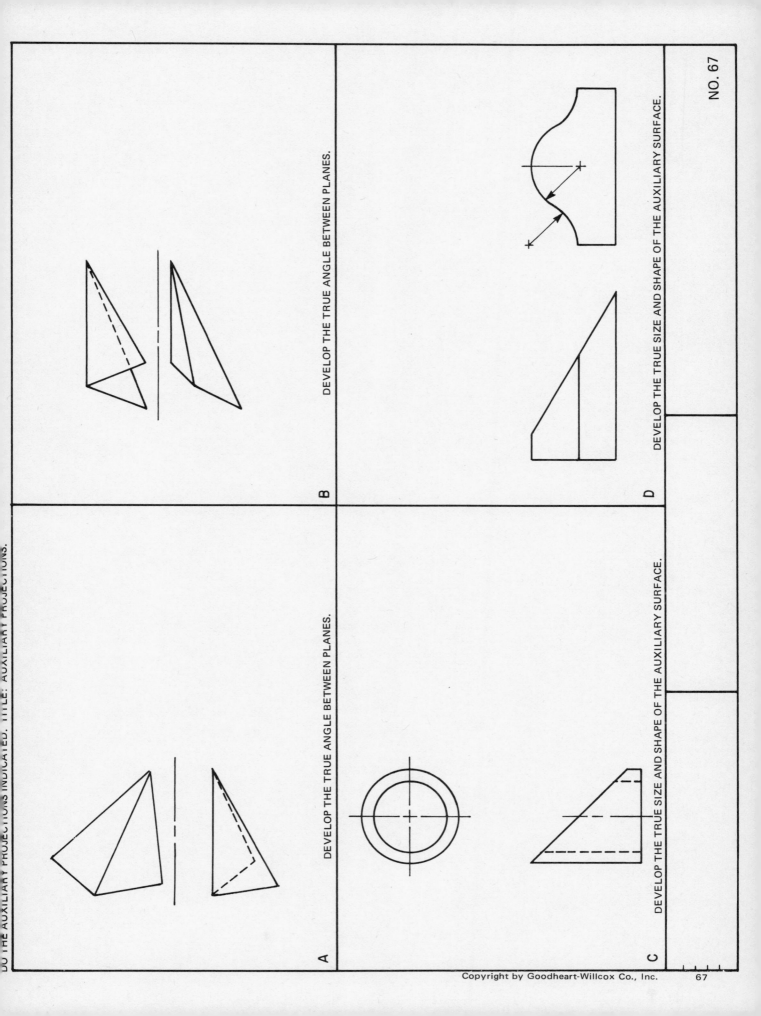

DO THE AUXILIARY PROJECTIONS INDICATED. TITLE: AUXILIARY PROJECTIONS.

A

DEVELOP THE TRUE ANGLE BETWEEN PLANES.

B

DEVELOP THE TRUE ANGLE BETWEEN PLANES.

C

DEVELOP THE TRUE SIZE AND SHAPE OF THE AUXILIARY SURFACE.

D

DEVELOP THE TRUE SIZE AND SHAPE OF THE AUXILIARY SURFACE.

NO. 67

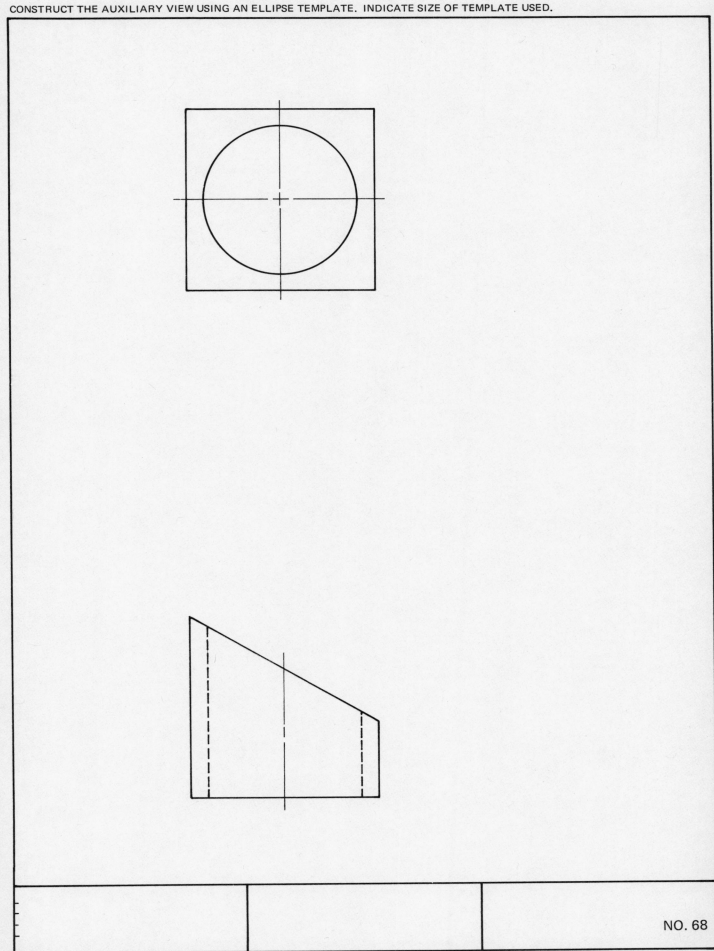

NO. 68

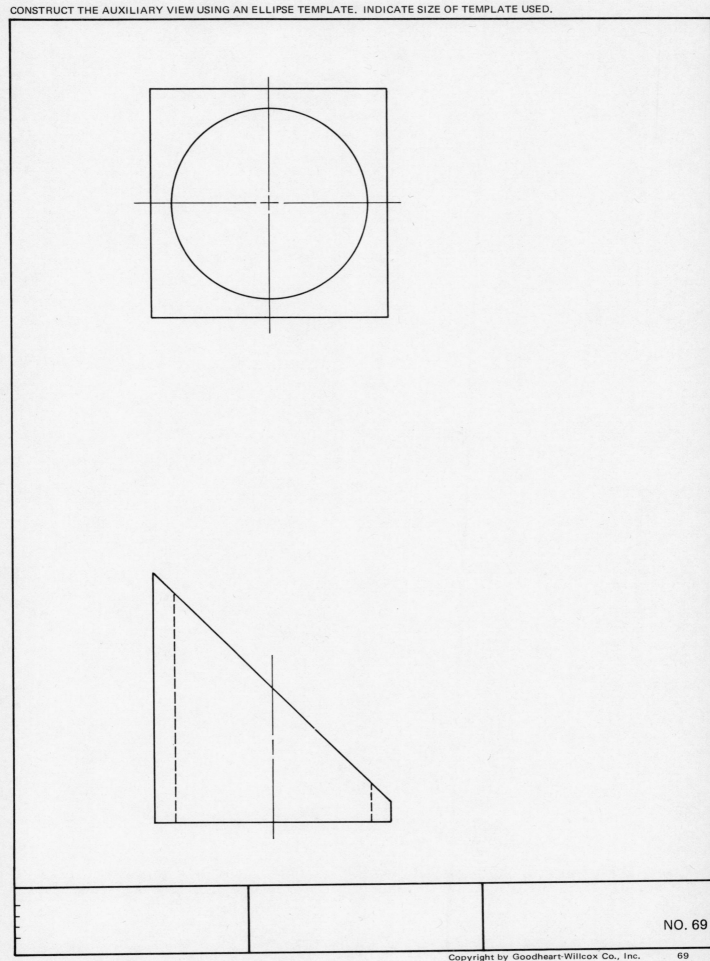

NO. 69

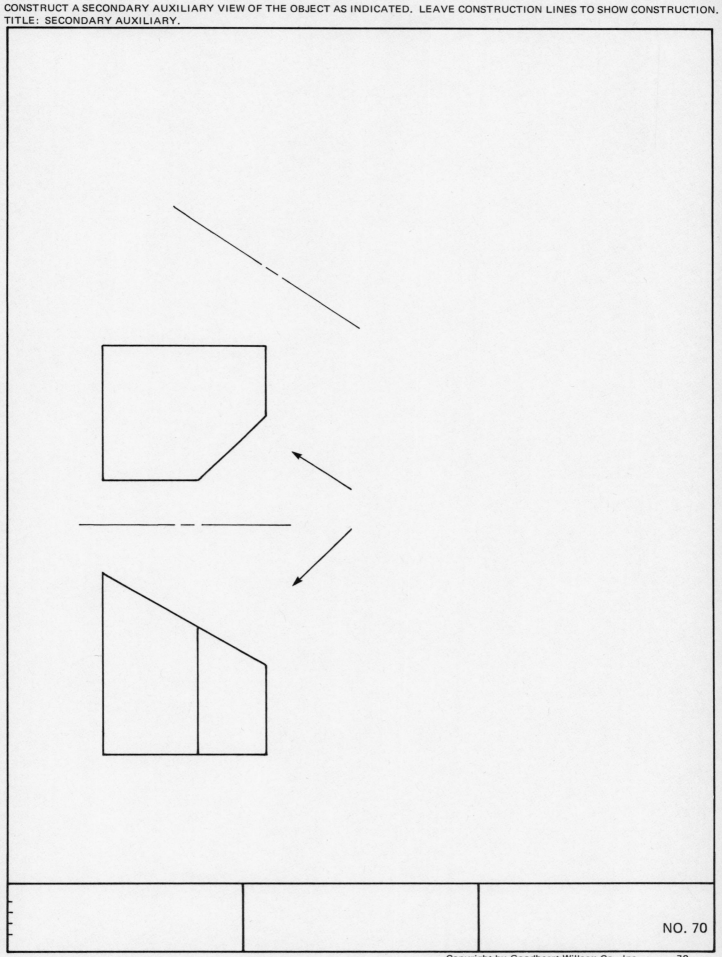

NO. 70

CONSTRUCT A SECONDARY AUXILIARY VIEW OF THE OBJECT AS INDICATED. LEAVE CONSTRUCTION LINES TO SHOW CONSTRUCTION. TITLE: SECONDARY AUXILIARY.

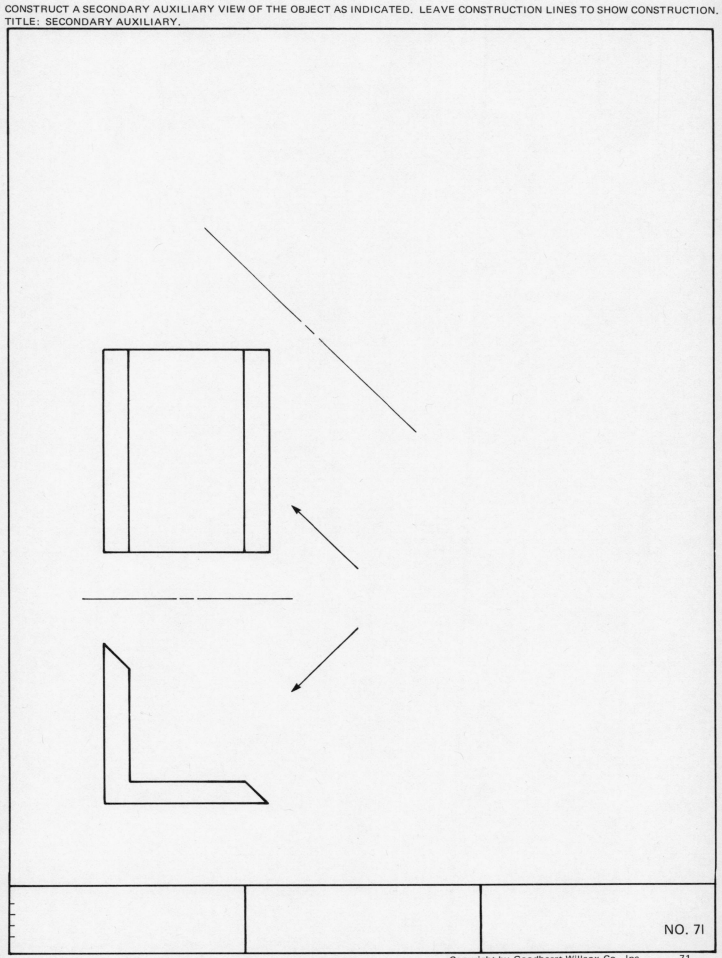

NO. 71

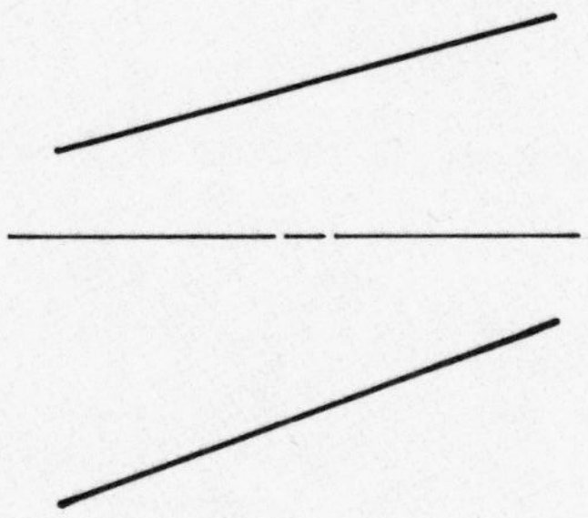

A FIND THE POINT VIEW OF THE LINE.

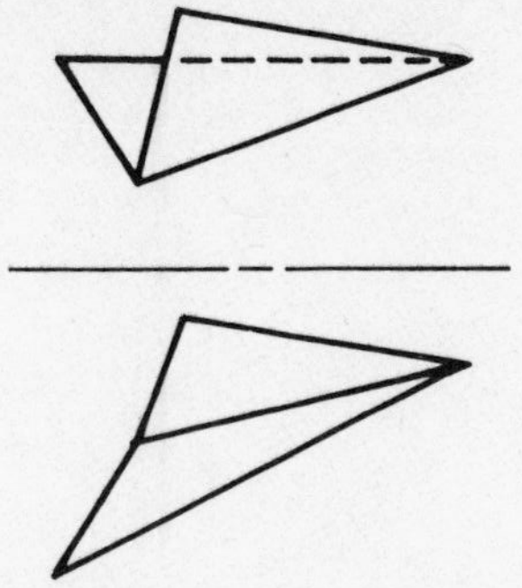

B FIND THE TRUE ANGLE BETWEEN THE OBLIQUE PLANES.

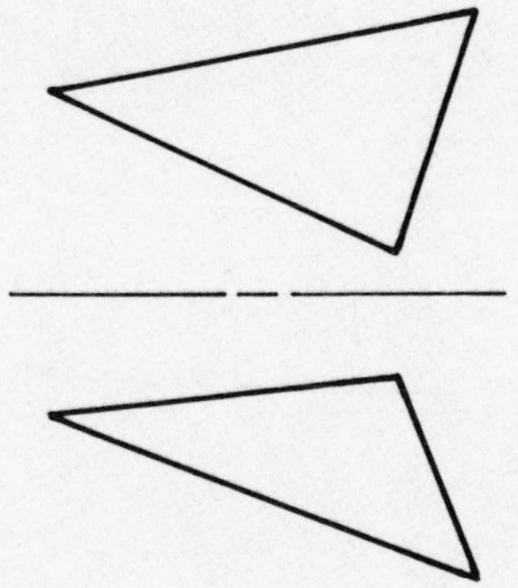

C CONSTRUCT THE TRUE SIZE AND SHAPE OF THE OBLIQUE SURFACE.

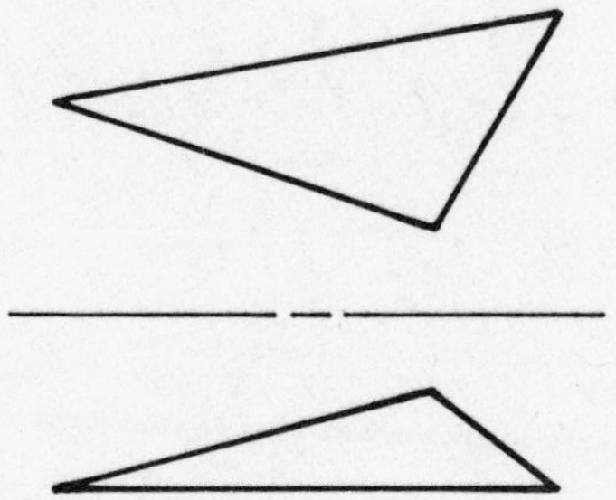

D CONSTRUCT THE TRUE SIZE AND SHAPE OF THE OBLIQUE SURFACE.

NO. 72

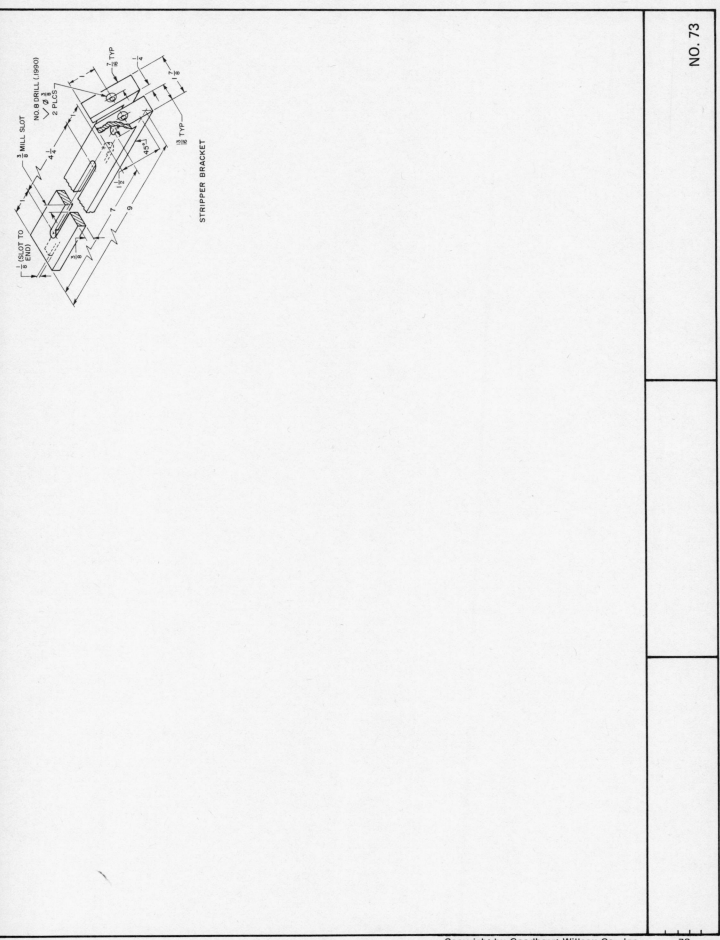

STRIPPER BRACKET

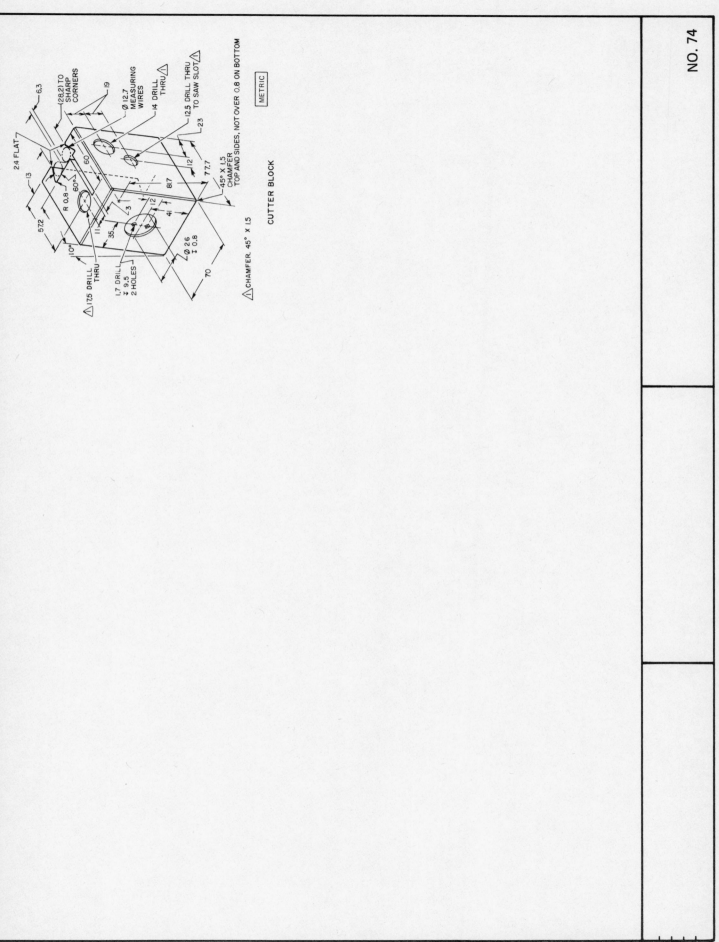

CUTTER BLOCK

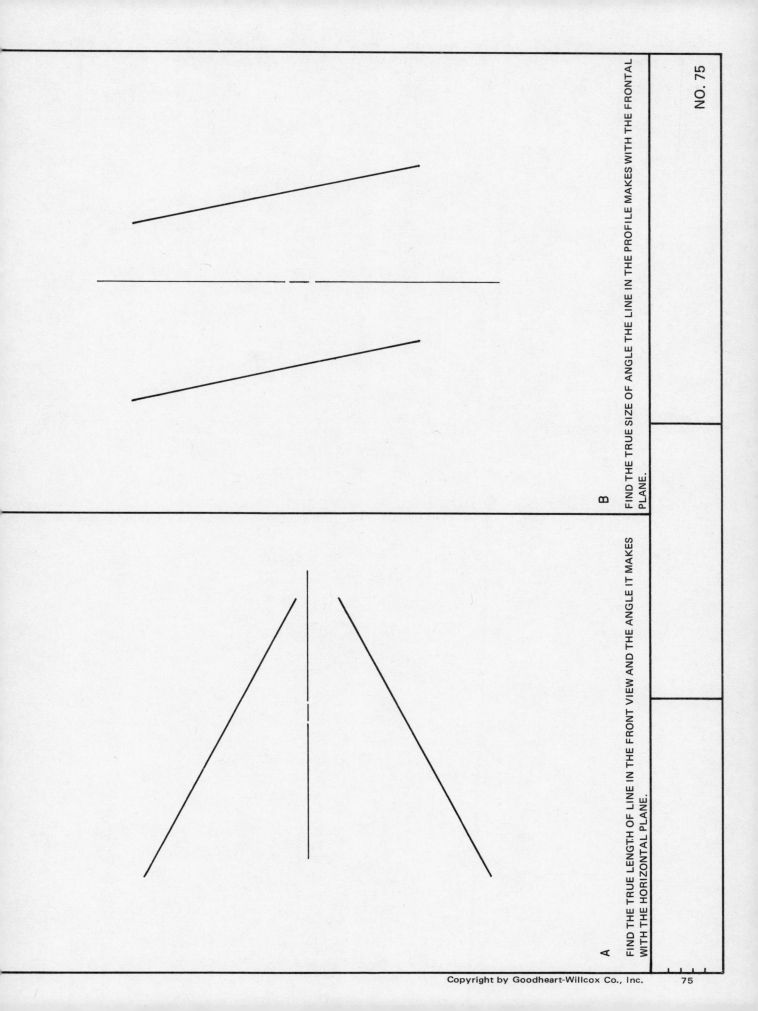

B

FIND THE TRUE SIZE OF ANGLE THE LINE IN THE PROFILE MAKES WITH THE FRONTAL PLANE.

A

FIND THE TRUE LENGTH OF LINE IN THE FRONT VIEW AND THE ANGLE IT MAKES WITH THE HORIZONTAL PLANE.

75

NO. 75

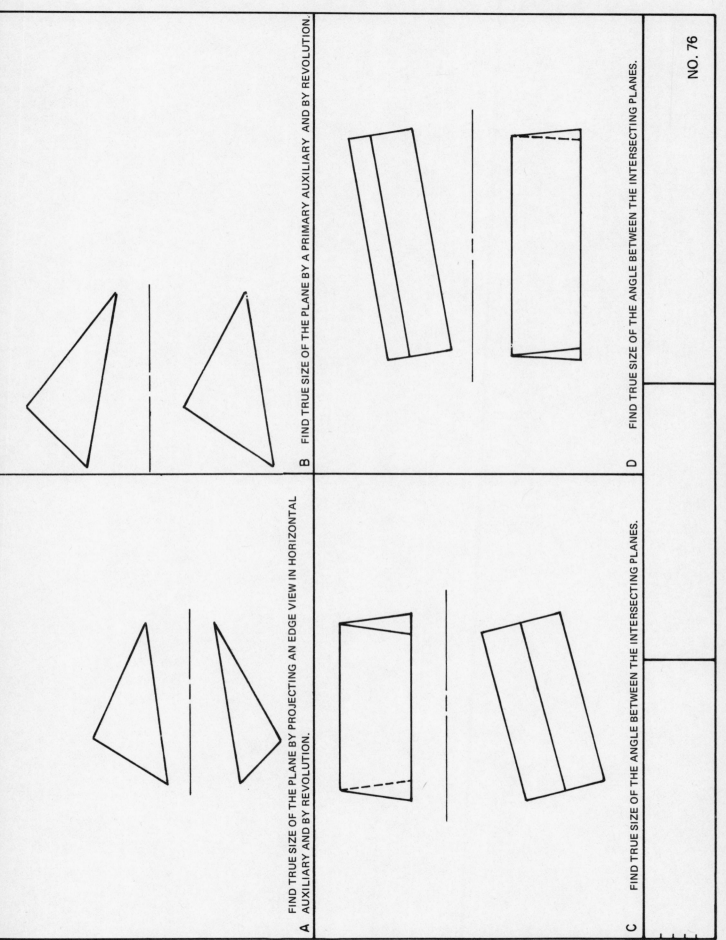

A FIND TRUE SIZE OF THE PLANE BY PROJECTING AN EDGE VIEW IN HORIZONTAL AUXILIARY AND BY REVOLUTION.

B FIND TRUE SIZE OF THE PLANE BY A PRIMARY AUXILIARY AND BY REVOLUTION.

C FIND TRUE SIZE OF THE ANGLE BETWEEN THE INTERSECTING PLANES.

D FIND TRUE SIZE OF THE ANGLE BETWEEN THE INTERSECTING PLANES.

NO. 76

USE THE REVOLUTION METHOD FOR SOLVING THE PROBLEMS AS INDICATED. TITLE: REVOLUTIONS.

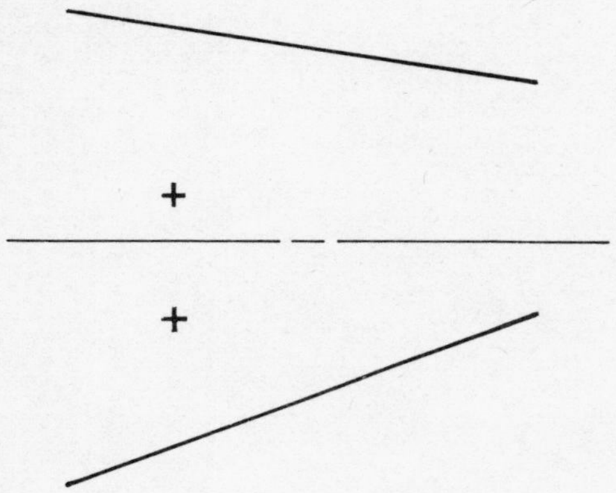

A

FIND THE PATH OF REVOLUTION OF THE POINT ABOUT THE LINE.

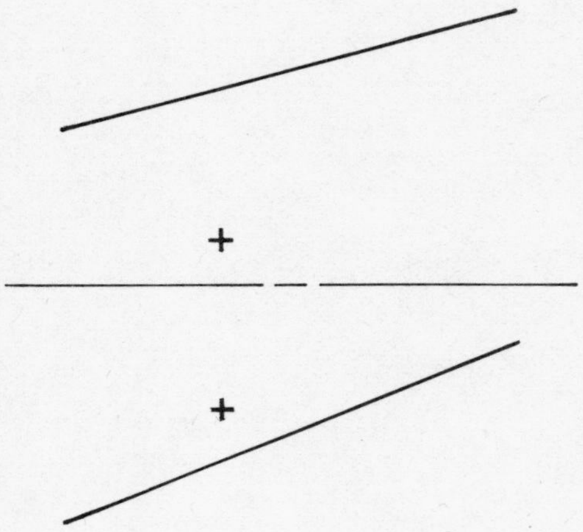

B

FIND THE PATH OF REVOLUTION OF THE POINT ABOUT THE LINE.

NO. 77

DRAW PRIMARY REVOLUTIONS OF THE OBJECT ABOUT THE AXES PERPENDICULAR TO THE PRINCIPAL PLANES AS INDICATED.
TITLE: PRIMARY REVOLUTIONS.

A

.12

1.50

1.25

.75

B   REVOLVE ABOUT AXIS OF HORIZONTAL PLANE.

C   REVOLVE ABOUT AXIS OF FRONTAL PLANE.

D   REVOLVE ABOUT AXIS OF PROFILE PLANE.

NO. 78

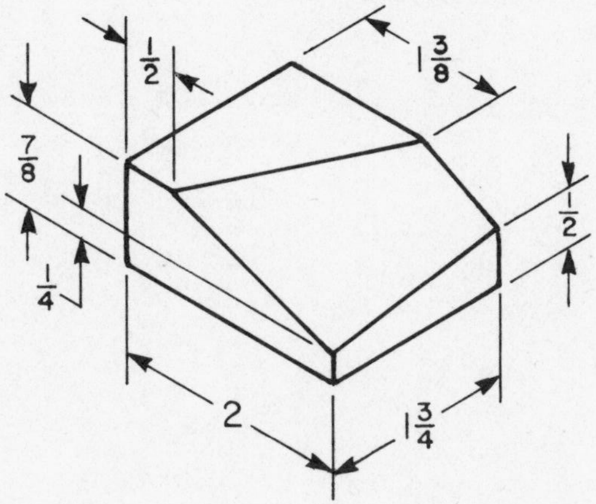

A

B   NORMAL POSITION.

C   REVOLVE ON AXIS OF HORIZONTAL PLANE.

D   REVOLVE ON AXIS OF FRONTAL PLANE. INDICATE THE TRUE SIZE SURFACE.

NO. 79

79

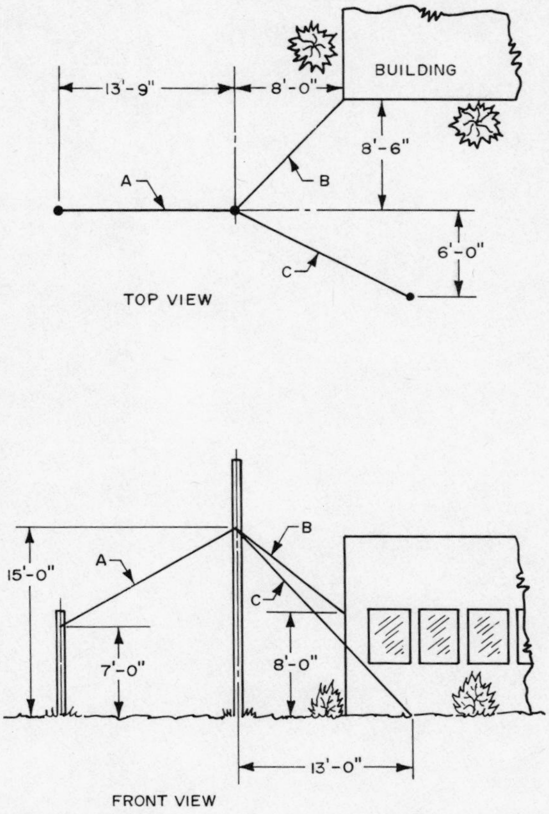

BUILDING

13'-9"

8'-0"

8'-6"

6'-0"

A

B

C

TOP VIEW

15'-0"

A

B

C

7'-0"

8'-0"

13'-0"

FRONT VIEW

A

B

C

NO. 80

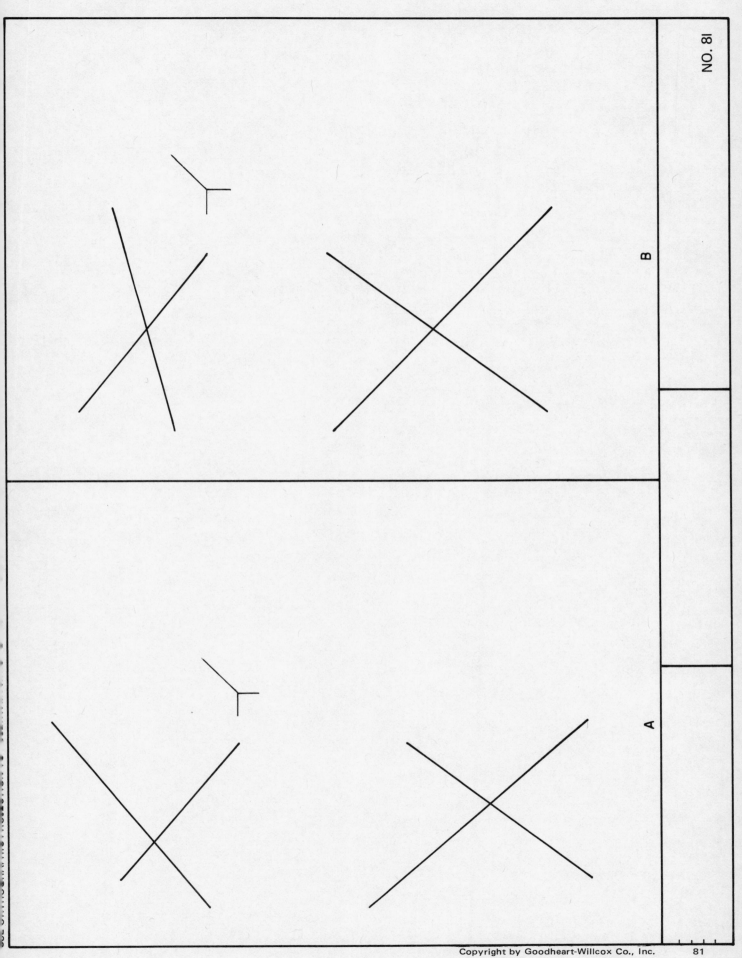

B

A

Ⓐ

Ⓑ

NO. 82

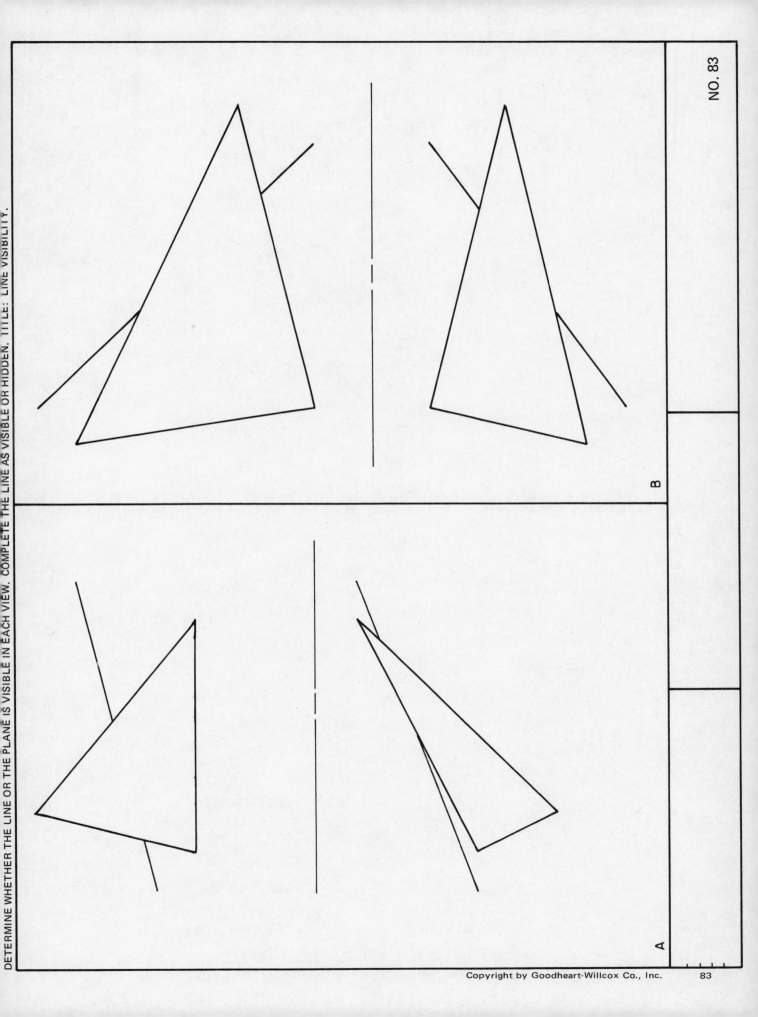

A

B

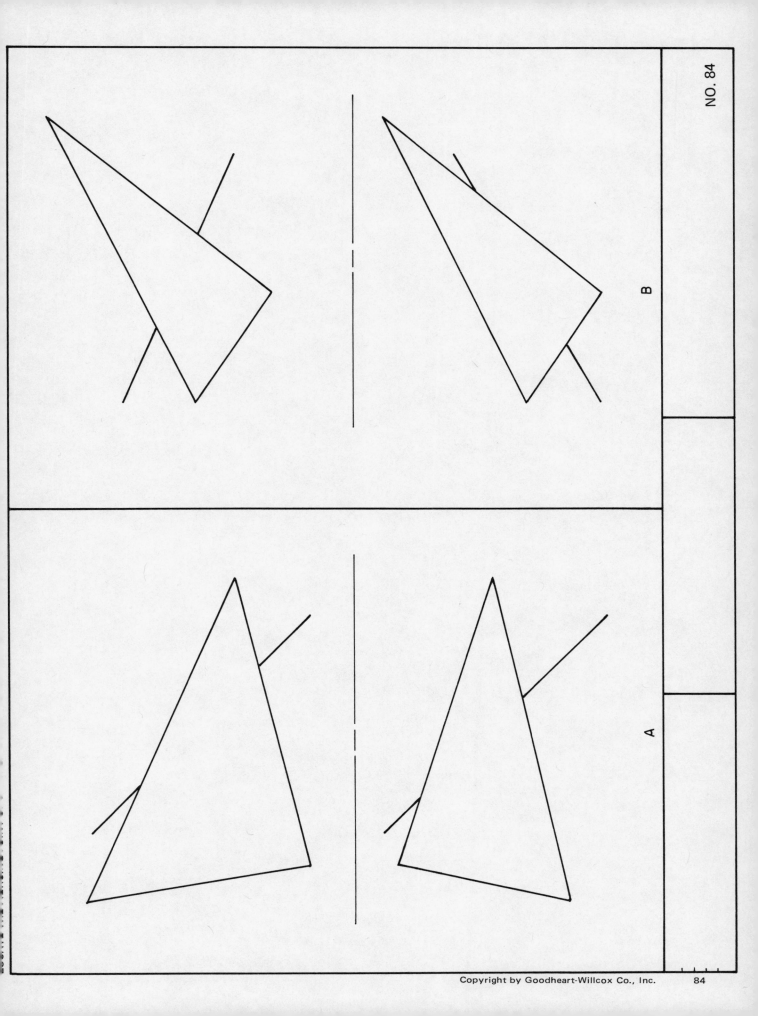

B

A

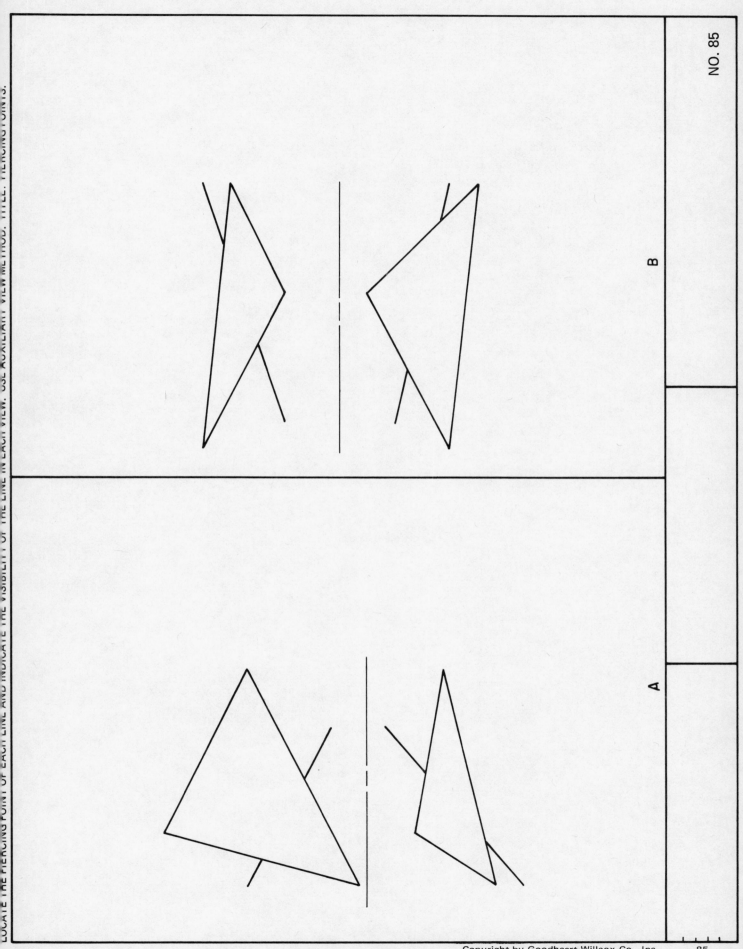

A

B

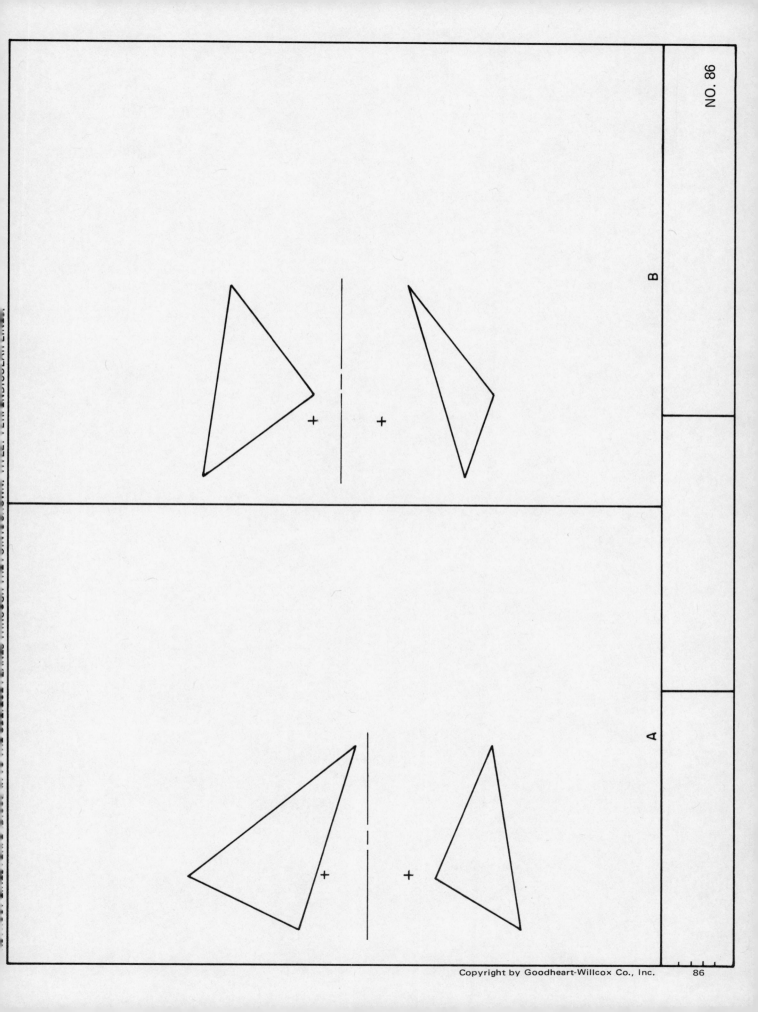

B

A

TITLE: INTERSECTION OF PLANES.

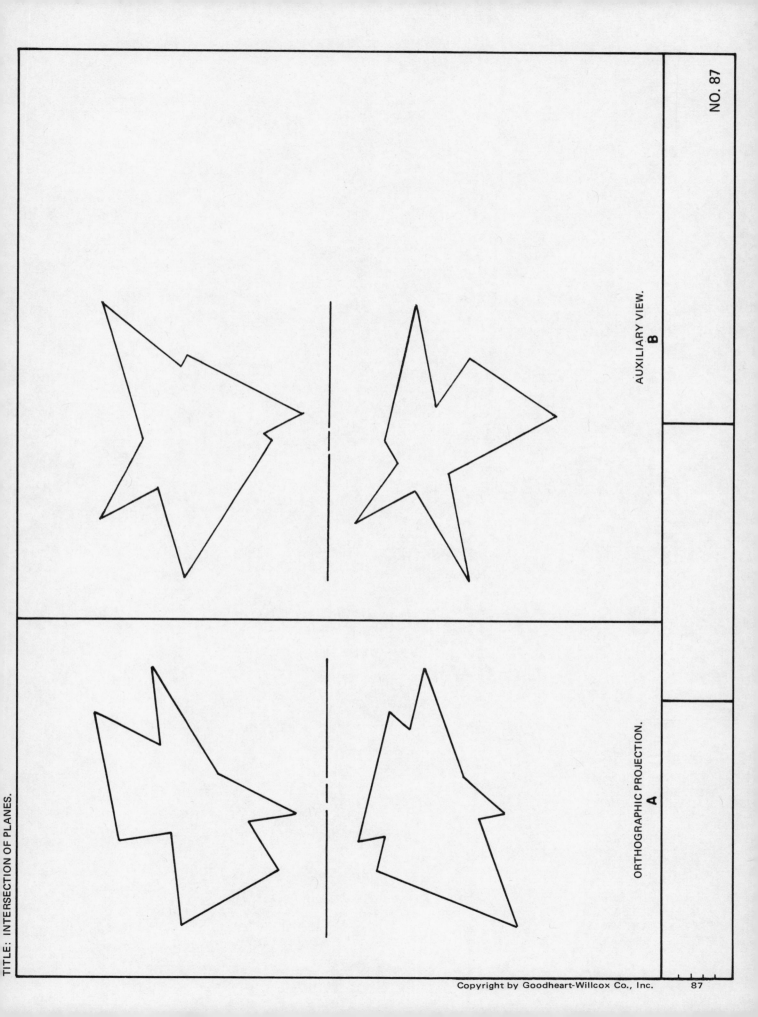

ORTHOGRAPHIC PROJECTION.
A

AUXILIARY VIEW.
B

NO. 88

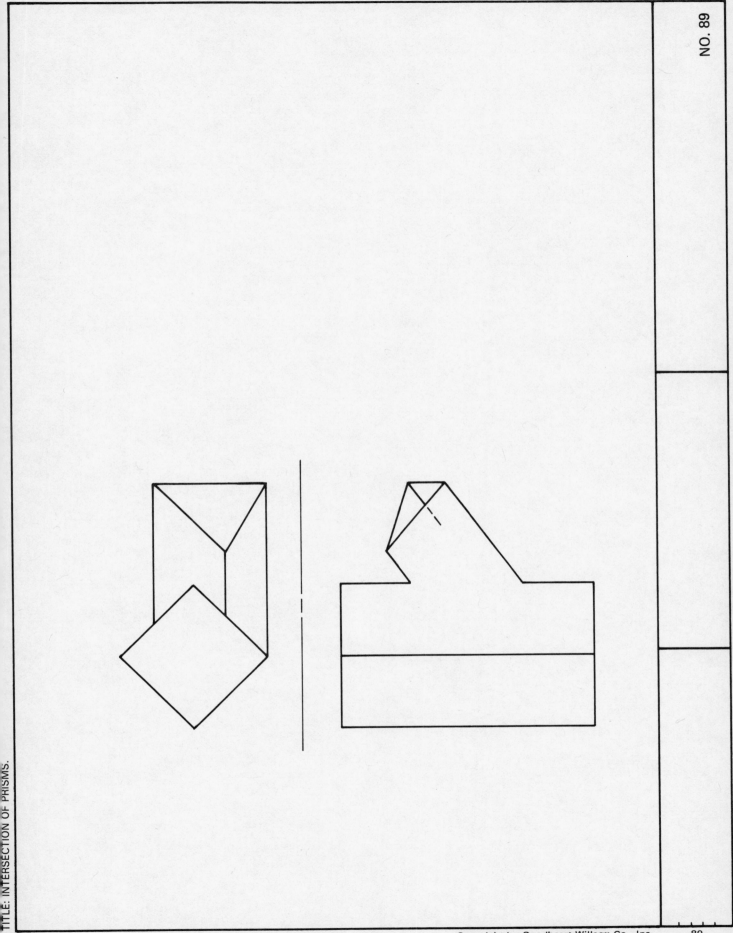

TITLE: INTERSECTION OF PRISMS.

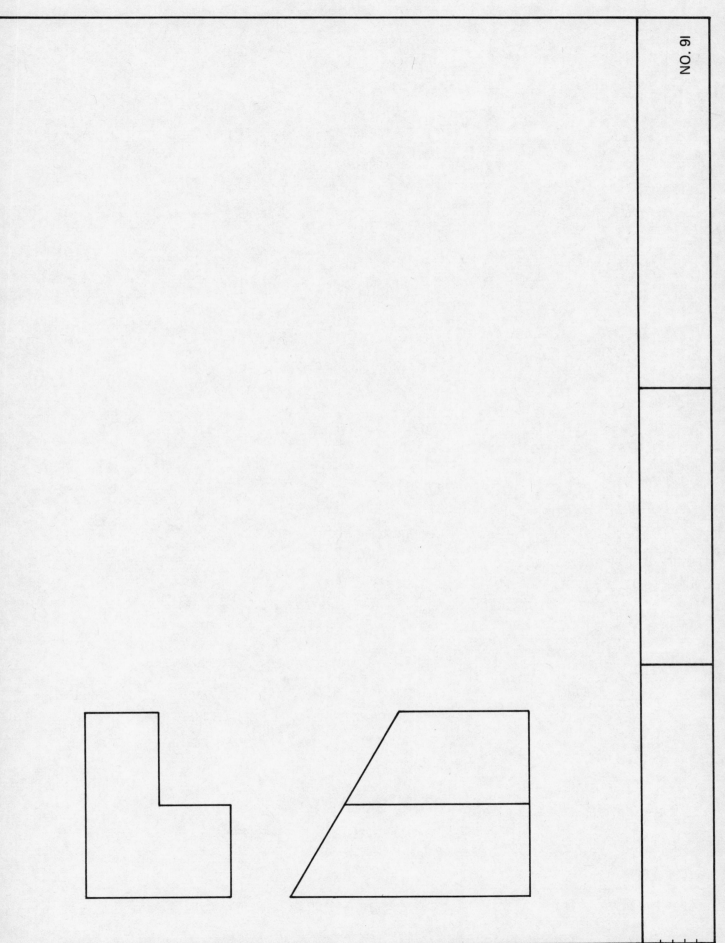

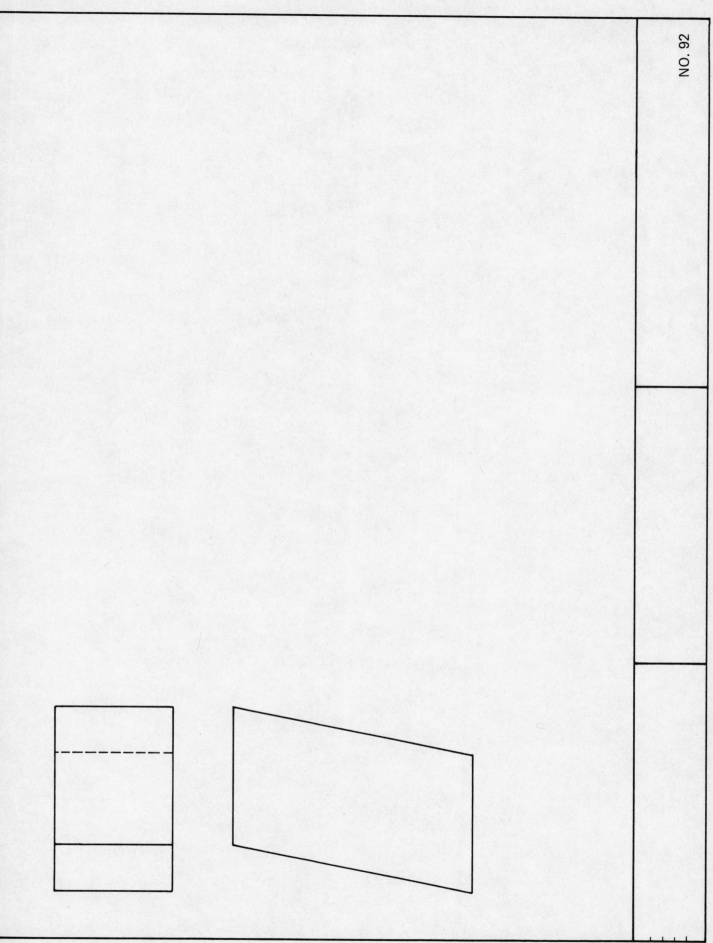

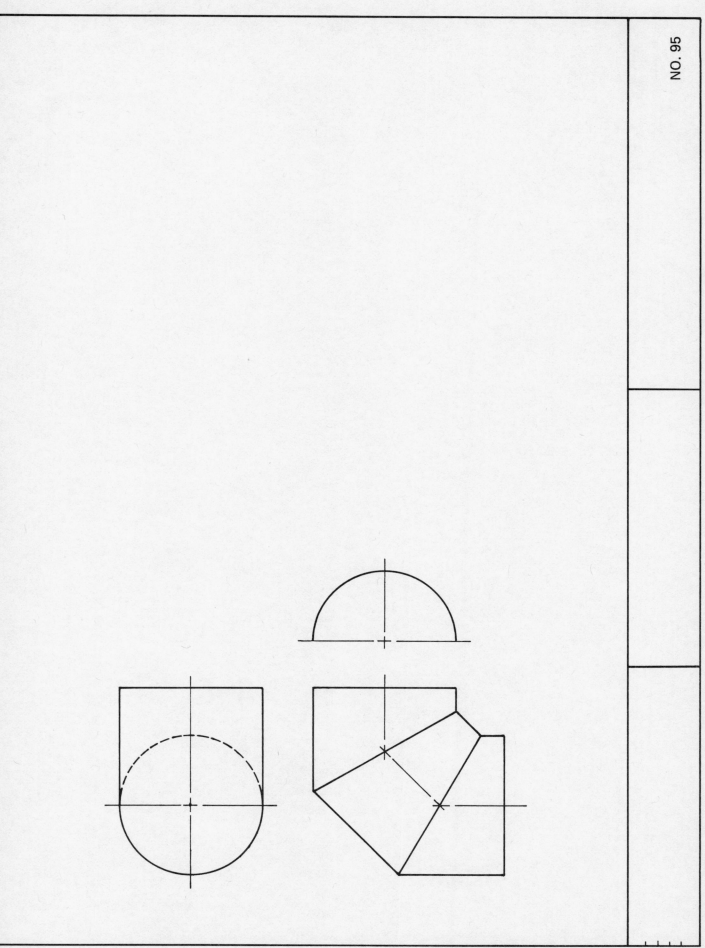

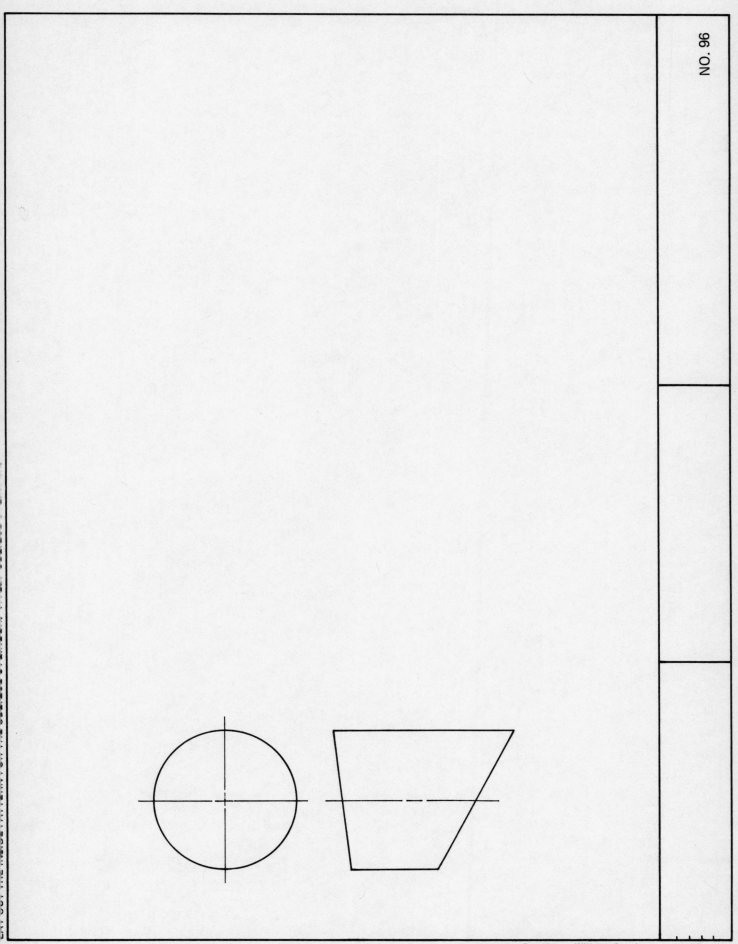

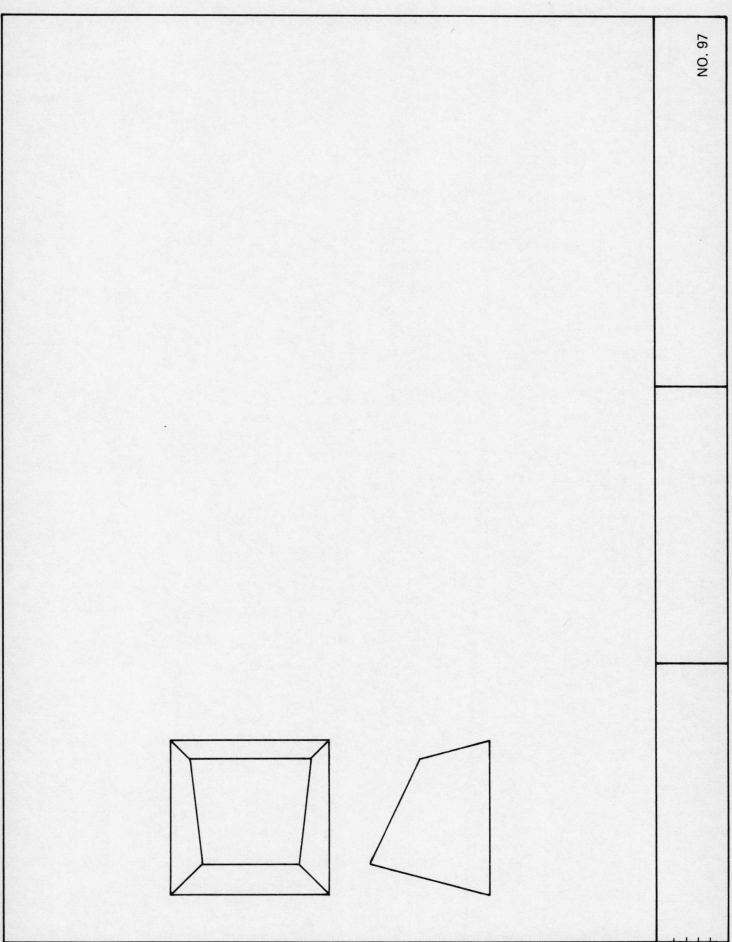

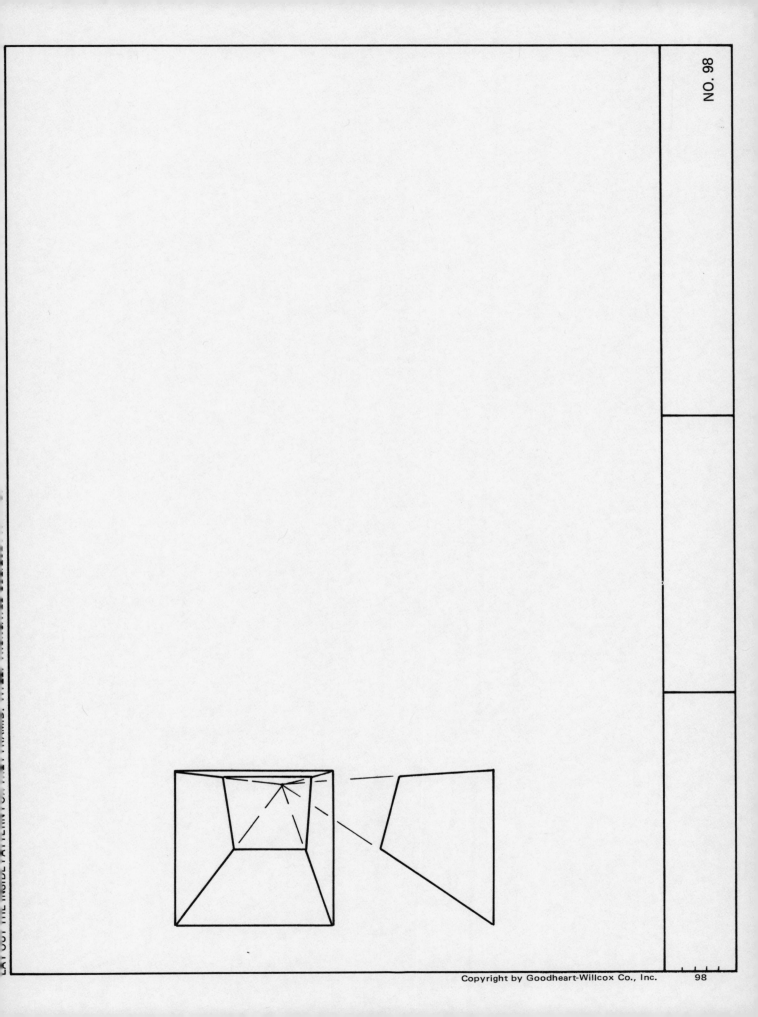

LAY OUT THE INSIDE PATTERN FOR THE CONE. TITLE: CONE DEVELOPMENT.

NO. 100

100

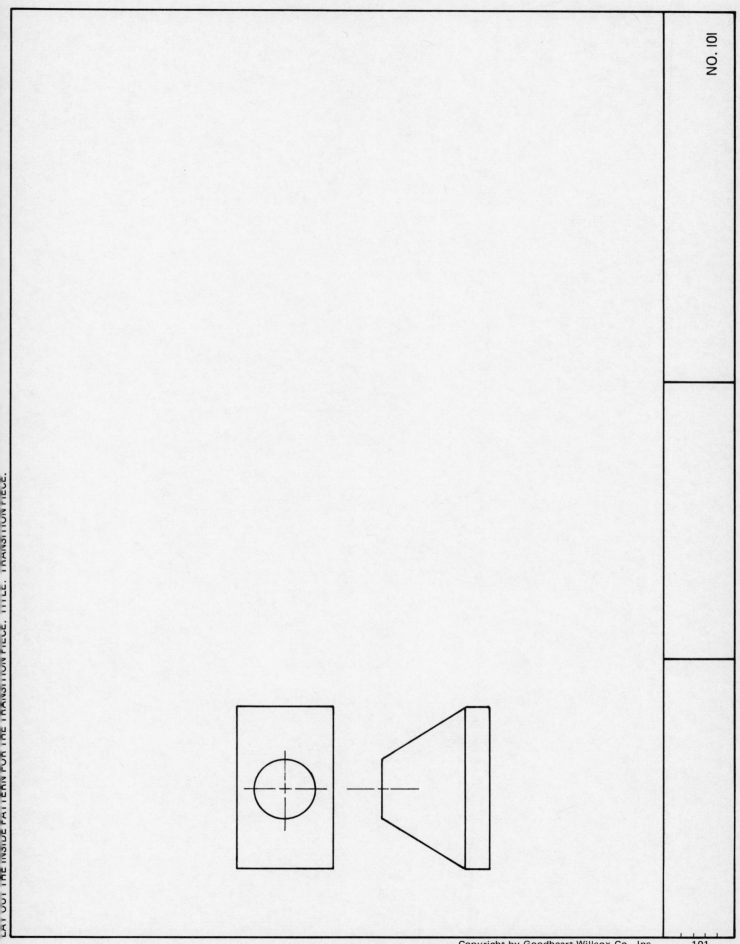

LAY OUT THE INSIDE PATTERN FOR THE TRANSITION PIECE. TITLE: TRANSITION PIECE.

DRAW A REGULAR HEXAGONAL AND SQUARE HEAD BOLT AND NUT OF THE FOLLOWING SPECIFICATION: NOMINAL SIZE 1/2 INCH, 3 INCHES LONG AND THREAD LENGTH 2 INCHES. SHOW THREAD IN SIMPLIFIED FORM. THE BOLT HEADS ARE TO BE SHOWN ON THE LEFT. DIMENSION THE THREAD WITH A NOTE.

NO. 102

SCHEMATIC THREAD REPRESENTATION AND DIMENSION THE PART. CHECK THE APPENDIX FOR DIMENSIONS NOT FURNISHED.

.250
.437
DRILL AND TAP
10 - 32UNF - 2B
.500
.375
Ø 2.50
.312
Ø .222
Ø .187
15°
30°
.062
.203
2.00

SPECIAL ADJUSTING SCREW

NO. 103

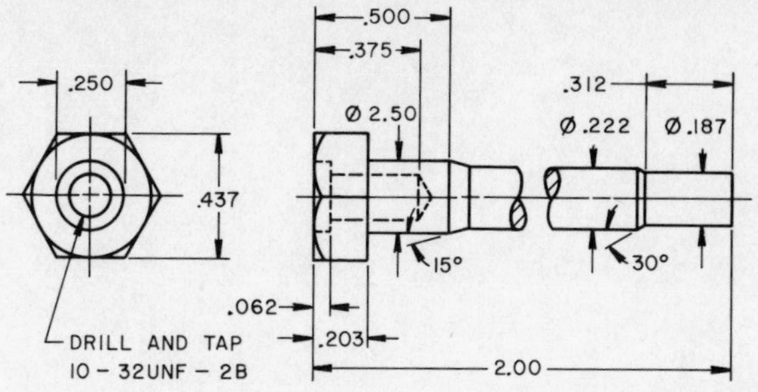

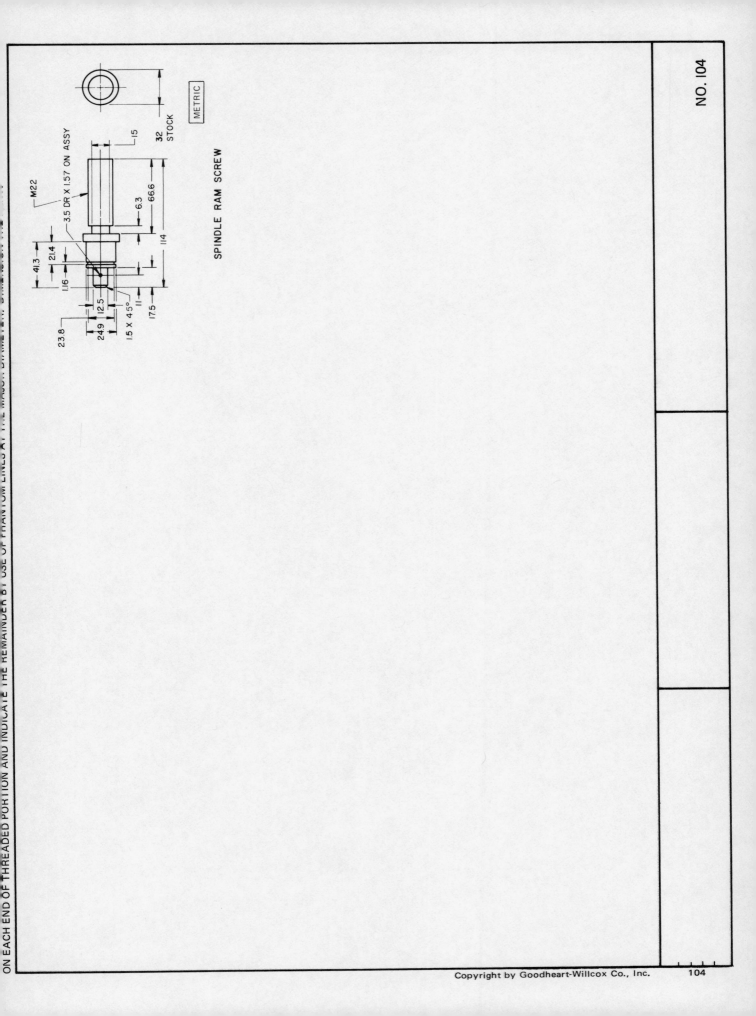

SPINDLE RAM SCREW

METRIC

M22

3.5 DR X 1.57 ON ASSY

15

32
STOCK

6.3

66.6

114

41.3

21.4

1.16

23.8

24.9

12.5

1.5 X 45°

11

17.5

ON EACH END OF THREADED PORTION AND INDICATE THE REMAINDER BY USE OF PHANTOM LINES AT THE MINOR DIAMETER. DIMENSION THE

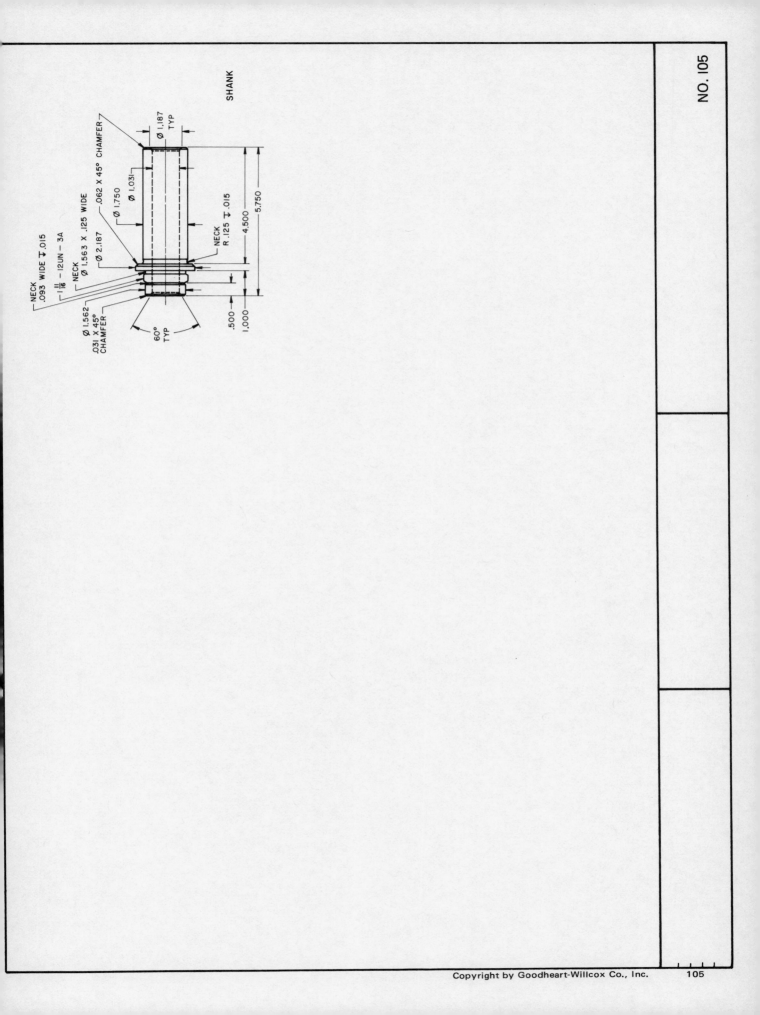

SHANK

NECK
.093 WIDE ⊤ .015

1 11/16 – 12UN – 3A

NECK
Ø 1.563 X .125 WIDE

Ø 2.187

.031 X 45°
CHAMFER

Ø 1.562

60°
TYP

.062 X 45° CHAMFER

Ø 1.750

Ø 1.031

NECK
R .125 ⊤ .015

Ø 1.187
TYP

5.750

4.500

1.000

.500

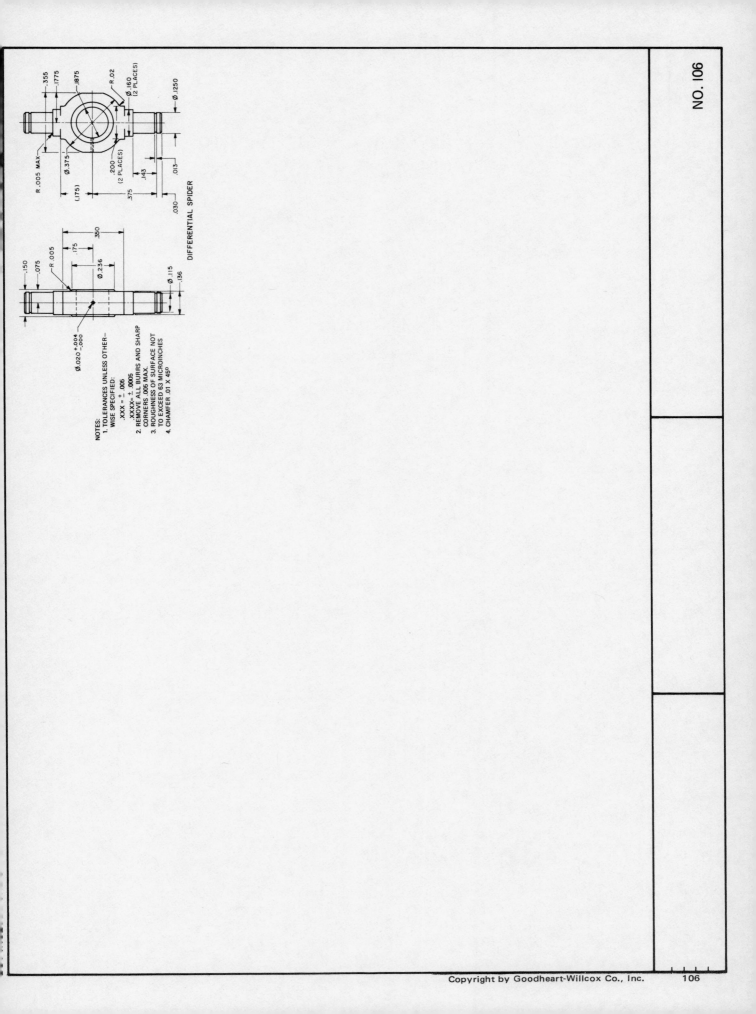

DIFFERENTIAL SPIDER

NOTES:
1. TOLERANCES UNLESS OTHER—
   WISE SPECIFIED:
   .XXX = ± .005
   .XXXX= ± .0005
2. REMOVE ALL BURRS AND SHARP
   CORNERS .005 MAX.
3. ROUGHNESS OF SURFACE NOT
   TO EXCEED 63 MICROINCHES
4. CHAMFER .01 X 45°

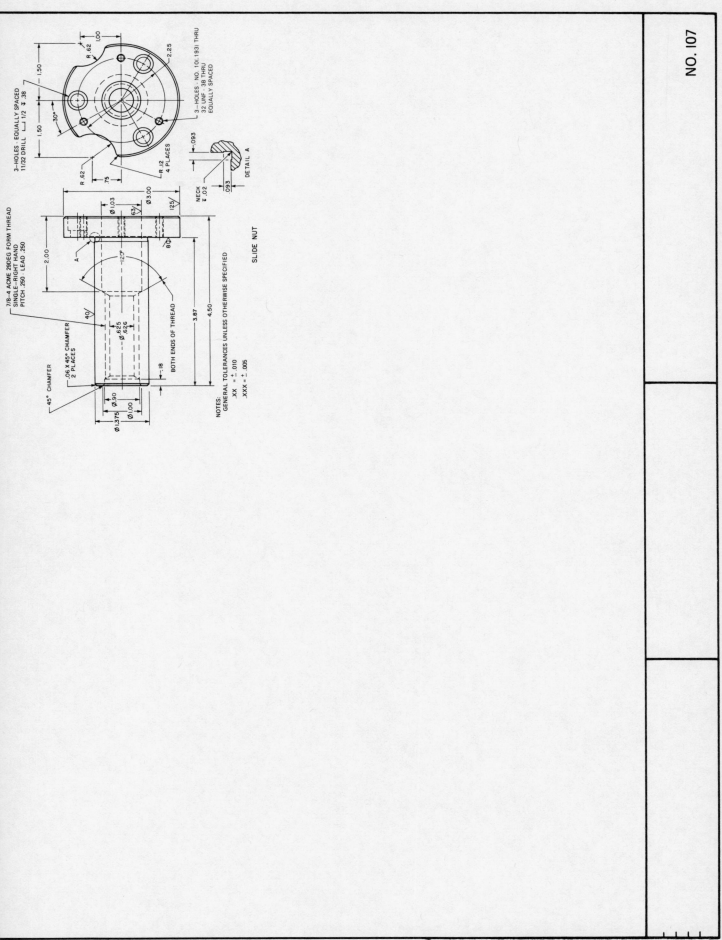

SLIDE NUT

3—HOLES - EQUALLY SPACED
11/32 DRILL ⌴ 1/2 ⩒ .38

3—HOLES - NO. 10(.193) THRU
32 UNF - 3B THRU
EQUALLY SPACED

R .62

1.00

2.25

1.50

1.50

30°

R .62

.75

R .12
4 PLACES

DETAIL A

NECK ⅀ .02

.093

.093

7/8—4 ACME 29DEG FORM THREAD
SINGLE—RIGHT HAND
PITCH .250 - LEAD .250

Ø 1.03

Ø 3.00

.63

125

80

2.00

.06 X 45° CHAMFER
2 PLACES

45° CHAMFER

A

40

120°

Ø .625
.626

BOTH ENDS OF THREAD

3.87

4.50

.18

Ø .90
Ø 1.00
Ø 1.375

NOTES:
GENERAL TOLERANCES UNLESS OTHERWISE SPECIFIED
.XX = ± .010
.XXX = ± .005

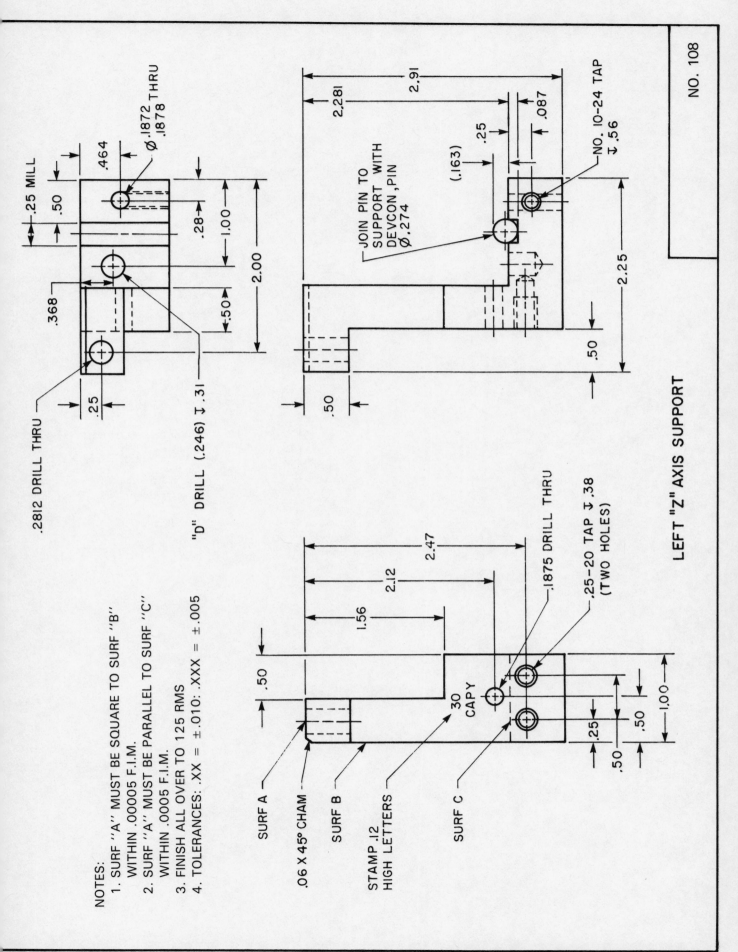

NOTES:
1. SURF "A" MUST BE SQUARE TO SURF "B" WITHIN .00005 F.I.M.
2. SURF "A" MUST BE PARALLEL TO SURF "C" WITHIN .0005 F.I.M.
3. FINISH ALL OVER TO 125 RMS
4. TOLERANCES: .XX = ±.010: .XXX = ±.005

.25 MILL
.50
.464
Ø .1872 THRU .1878

.28
1.00
.368
2.00
.50
.25
.2812 DRILL THRU
"D" DRILL (.246) ↧ .31

2.281
2.91
.087
.25
(.163)
JOIN PIN TO SUPPORT WITH DEVCON, PIN Ø .274
NO. 10-24 TAP ↧ .56
2.25
.50
.50

2.47
2.12
1.56
.50
.1875 DRILL THRU
.25-20 TAP ↧ .38 (TWO HOLES)
1.00
.25
.50
.50
30 CAPY
.06 x 45° CHAM
SURF A
SURF B
STAMP .12 HIGH LETTERS
SURF C

LEFT "Z" AXIS SUPPORT

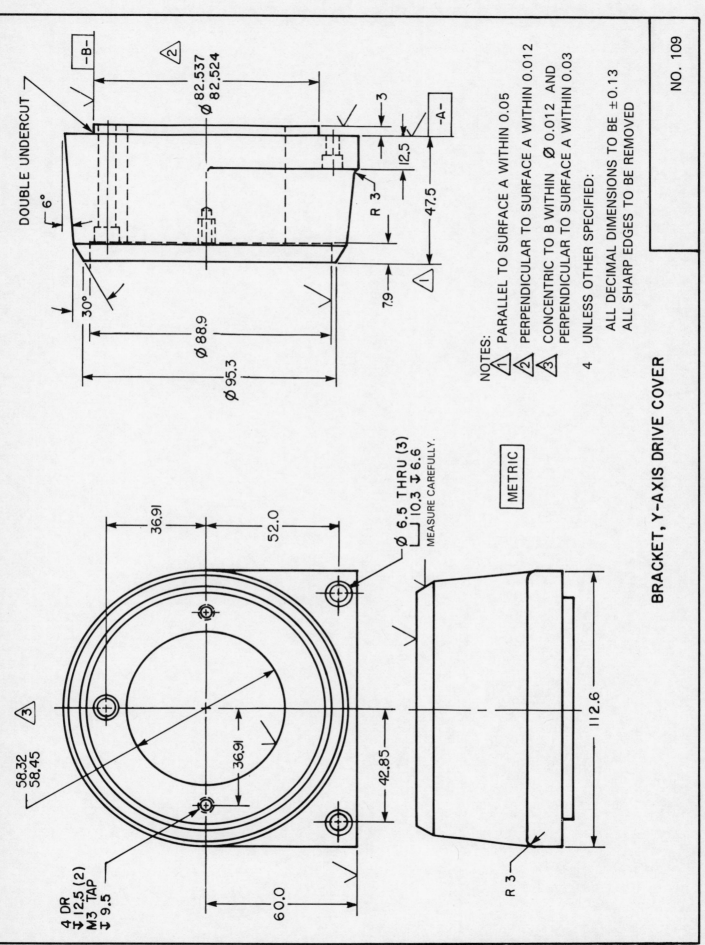

DOUBLE UNDERCUT

6°

30°

Ø 82.537
82.524

-B-

-A-

3

12,5

47,5

7,9

R 3

Ø 88.9

Ø 95.3

NOTES:
△1 PARALLEL TO SURFACE A WITHIN 0.05
△2 PERPENDICULAR TO SURFACE A WITHIN 0.012
△3 CONCENTRIC TO B WITHIN Ø 0.012 AND
   PERPENDICULAR TO SURFACE A WITHIN 0.03

4 UNLESS OTHER SPECIFIED:
   ALL DECIMAL DIMENSIONS TO BE ±0.13
   ALL SHARP EDGES TO BE REMOVED

36,91

52.0

Ø 6.5 THRU (3)
⌴ 10.3 ⌴ 6.6
MEASURE CAREFULLY.

METRIC

58,32
58,45

36,91

42,85

60.0

112.6

R 3

4 DR
⌴ 12.5 (2)
M3 TAP
⌴ 9.5

BRACKET, Y-AXIS DRIVE COVER

NO. 109

MODIFIED UNIFORM MOTION (USE ARC OF ONE-QUARTER OF THE RISE TO MODIFY THE UNIFORM MOTION IN THE DISPLACEMENT DIAGRAM) WITH A RISE OF .687''. THE CAM ROTATES CLOCKWISE.

CAM LAYOUT

DISPLACEMENT DIAGRAM

NO. 110

10°. THE CAM OPERATES AT MODERATE SPEED. YOU ARE TO SELECT THE APPROPRIATE CAM MOTION, SIZE OF BASE CIRCLE AND TYPE OF CAM FOLLOWER AND MAKE A FULL-SIZE WORKING DRAWING OF THE DISPLACEMENT DIAGRAM AND CAM.

NO. III

WIDTH 1.00, FACE WIDTH .50 AND KEYWAY $\frac{1}{8}$ × $\frac{1}{16}$ INCH. COMPUTE THE VALUES FOR THE PITCH DIAMETER, CIRCULAR THICKNESS AND WHOLE DEPTH, AND INCLUDE THESE IN A TABLE ON THE DRAWING. ONE VIEW SHOULD BE A SECTIONAL VIEW.

NO. 1I2

LENGTH 1.25, MOUNTING DISTANCE 1.875 INCHES, HUB DIAMETER 2.125 INCHES AND $\frac{1}{8} \times \frac{1}{16}$ KEYWAY. MAKE ONE VIEW A SECTIONAL VIEW FOR CLARITY AND COMPUTE THE VALUES FOR PITCH DIAMETER, CIRCULAR PITCH AND WHOLE DEPTH, ADDENDUM, DEDENDUM, AND INCLUDE THESE IN A TABLE ON THE DRAWING.

NO. 113

MACHINIST'S HANDBOOK OR FROM A GEAR CATALOG. MAKE AN ASSEMBLY DRAWING OF THE GEARS AND PLACE THE NECESSARY DIMENSIONS AND SPECIFICATIONS ON THE DRAWING.

NO. 114

114

MAKE A DETAILED WORKING DRAWING OF THE PART SHOWN. USE AN A-SIZE SHEET. CHANGE ALL DIMENSIONS TO DECIMAL LIMIT DIMEN-
SIONS WITH THE FOLLOWING TOLERANCES: .XXX = ±.003; .XX = ±.010. DELETE ALL UNNECESSARY DIMENSIONS. INDICATE BY USE OF
SYMBOLS ALL FLAT SURFACES AS 125 MICROINCHES AND ALL BORED AND COUNTEBORED HOLES AS 63 MICROINCHES IN TEXTURE.

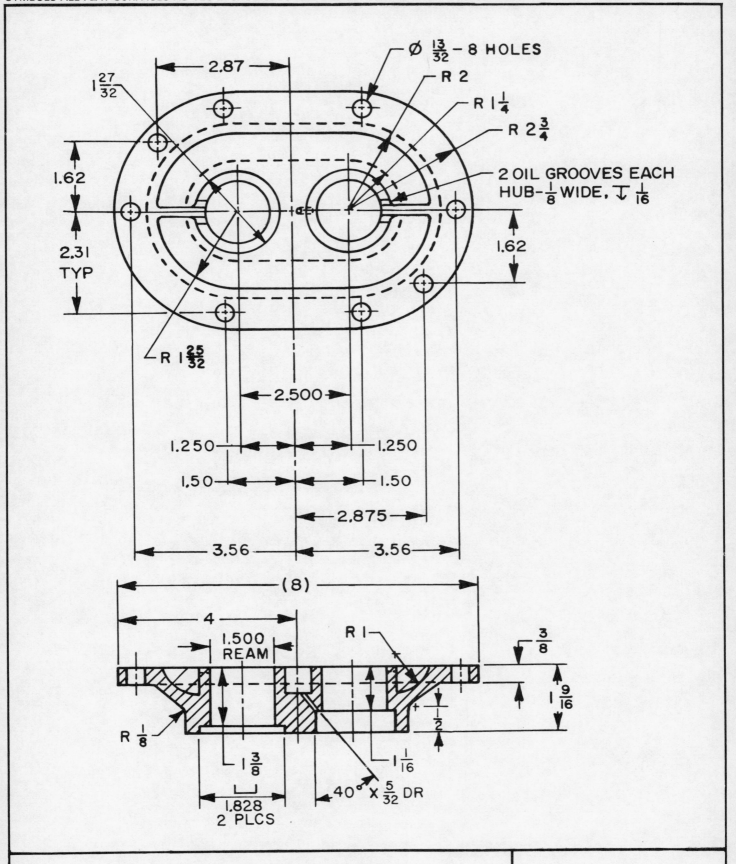

GEAR COVER PLATE

NO. 115

DIMENSION FOR N/C MACHINING USING THE CHART IN THE APPENDIX OF THE TEXT FOR CALCULATING THE EQUALLY SPACED LOCATIONS. DIMENSION THE DRAWING USING GEOMETRIC FEATURE CONTROL SYMBOLS. USE A B-SIZE SHEET OF DRAWING PAPER.

.375 DR ⊥ .90
LOCATED WITHIN R .010
OF TRUE POSITION

63

Ø 5.50

Ø 6.524–6.526
FOR ⟨3⟩

.04 X 30° CHAM

.06 X 30° CHAM

BO 3.485
–3.487
GR 3.500
–3.510

R .03

15

.82
.81

7°

.380
.375

7°

R .06

.10

SEE A

R .03

.04

.05
.04

.12

1.10

3.340
3.330

.50

.04 X 45° CHAM

Ø 3.80

R .14

BO 4.315
–4.320
GR 4.3303
–4.3308
FOR ⟨2⟩

.06 X 30° CHAM

⊥ .04 RELIEF

R .04

.03   .04

DETAIL A

TR 1.365–1.370
GR 1.375–1.380

Ø TR 7.010
–7.015
Ø GR 6.996
–6.997

Ø 5.60

.438 FLAT BOTTOM DR 1.495 – ⊥ 1.500
24 HOLES EQUALLY SPACED FOR ⟨1⟩
LOC WITHIN .010 TRUE POS

.128 DR THRU TO BORE
LOCATE WITHIN R .01 OF
TRUE POSITION

(6.24)

2.26

PARTS LIST			
ITEM	PART NO	NAME	QUAN
⟨1⟩	11000063	SPRING	24
⟨2⟩	29012122	BRG	1
⟨3⟩	11330166	"O" RING	1

NO. 116

NOTES:
1. FINISH ALL OVER
2. ALL UNTOLERANCED DIMENSIONS ARE ±.02 EXCEPT CASTING, FABRICATION, FORGING AND HOLE DIAMETERS LISTED IN STANDARD TOLERANCE CHARTS.

HYDRAULIC DECHUCK PISTON

116

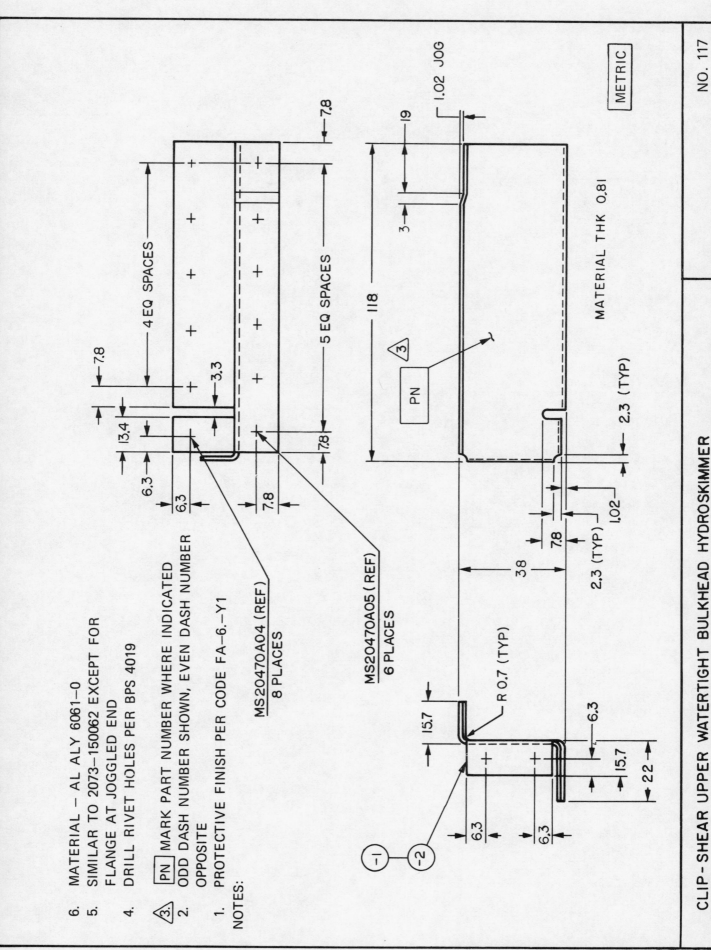

NOTES:

1. PROTECTIVE FINISH PER CODE FA–6.–Y1

2. ODD DASH NUMBER SHOWN, EVEN DASH NUMBER OPPOSITE

△3. PN MARK PART NUMBER WHERE INDICATED

4. DRILL RIVET HOLES PER BPS 4019

5. SIMILAR TO 2073–150062 EXCEPT FOR FLANGE AT JOGGLED END

6. MATERIAL – AL ALY 6061–0

MS20470A04 (REF)
8 PLACES

MS20470A05 (REF)
6 PLACES

4 EQ SPACES

5 EQ SPACES

7.8

3.3

13.4

6.3

6.3

7.8

7.8

1.02 JOG

19

3

118

PN

△3

MATERIAL THK 0.81

2.3 (TYP)

1.02

7.8

2.3 (TYP)

38

R 0.7 (TYP)

15.7

15.7

22

6.3

6.3

6.3

–1

–2

CLIP – SHEAR UPPER WATERTIGHT BULKHEAD HYDROSKIMMER

METRIC

NO. 117

MAKE AN ASSEMBLY WORKING DRAWING OF THE PARTS SHOWN. ADD BILATERAL TOLERANCES TO THOSE PARTS REQUIRING FITS AS INDICATED IN THE NOTES. USE AN A-SIZE DRAWING SHEET.

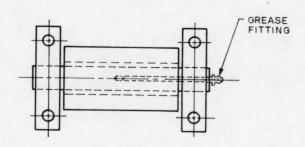

GREASE FITTING

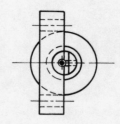

ASSEMBLY VIEWS

ROLLER – C.R.S.

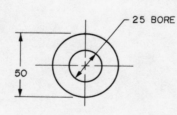

25 BORE

50

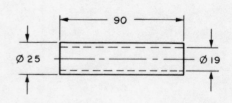

90

Ø 25   Ø 19

BUSHING – BRONZE

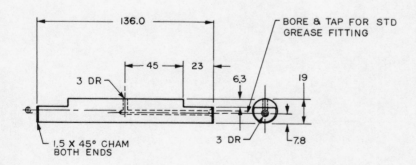

136.0

45   23

3 DR

BORE & TAP FOR STD GREASE FITTING

6.3   19

3 DR

7.8

1.5 X 45° CHAM BOTH ENDS

SHAFT – C.R.S.

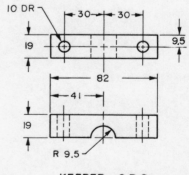

10 DR   30   30

19   9.5

82

41

19

R 9.5

KEEPER – C.R.S.

NOTES:
2.   FINISH 125/ ALL OVER
1.   BUSHING TO BE A LIGHT DRIVE FIT
     IN ROLLER & RUNNING FIT ON SHAFT

METRIC

ROLLER FOR BRICK ELEVATOR   NO. 118

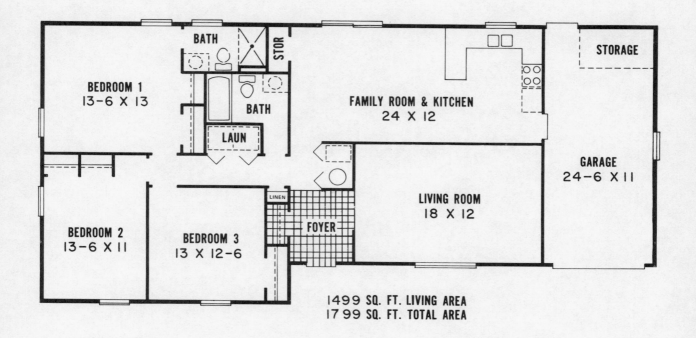

PREPARE DRAWINGS AS INDICATED IN THE FOLLOWING PROBLEMS FOR THE HOUSE ILLUSTRATED IN THE PRESENTATION DRAWING SHOWN. PROBLEMS ARE BEST SUITED FOR B- OR C-SIZE SHEETS.

SHEET NO.1

PREPARE A SCALED DRAWING OF THE FLOOR PLAN OF THE HOUSE. INCLUDE ALL NECESSARY DIMENSIONS AND NOTES. DO NOT INCLUDE ELECTRICAL PLAN.

SHEET NO. 2

MAKE A FOOTING AND FOUNDATION PLAN FOR THE HOUSE. THERE IS NO BASEMENT AND THE FOUNDATION IS A 36 INCH STEM WALL ON A FOOTING. ON THE SAME SHEET, PREPARE A DETAIL SECTION OF THE FOUNDATION SHOWING ANY BEAMS AND PIERS NECESSARY IN THE FOUNDATION TO SUPPORT THE FLOOR AND INTERIOR WALLS.

SHEET NO. 3

TRACE THE FLOOR PLAN PREPARED IN THE PROBLEM AND ADD AN ELECTRICAL WIRING PLAN FOR THE HOUSE. CHECK THE LOCAL ELECTRICAL CODE, IF ONE IS AVAILABLE, FOR THE REQUIREMENTS ON SPACING WALL OUTLETS. SHOW LINES TO SWITCHES ON ALL OUTLETS CONTROLLED BY SWITCHES.

SHEET NO. 4

MAKE A FRONT ELEVATION FOR THE HOUSE. ADD DIMENSIONS AND NOTES WHERE NECESSARY.

SHEET NO. 5

DRAW A SIDE ELEVATION FOR THE HOUSE AND DIMENSION.

SHEET NO. 6

MAKE A WALL SECTION DRAWING TO SHOW DETAILS OF CONSTRUCTION. ADD DIMENSIONS AND NOTES WHERE NECESSARY.

NO. 119

DRAW AND LABEL THE FOLLOWING COMPONENT SYMBOLS. USE A TEMPLATE OR DRAW EACH COMPONENT TO THE SAME RELATIVE SIZE.

A. BATTERY, 9 VOLTS, BTI.

B. SWITCH, SINGLE-POLE THROW, SI.

C. AMMETER, MI.

D. RESISTOR, 4700 OHMS, RI.

E. LAMP, INCANDESCENT, DIAL LAMP, DSI.

NO. 120

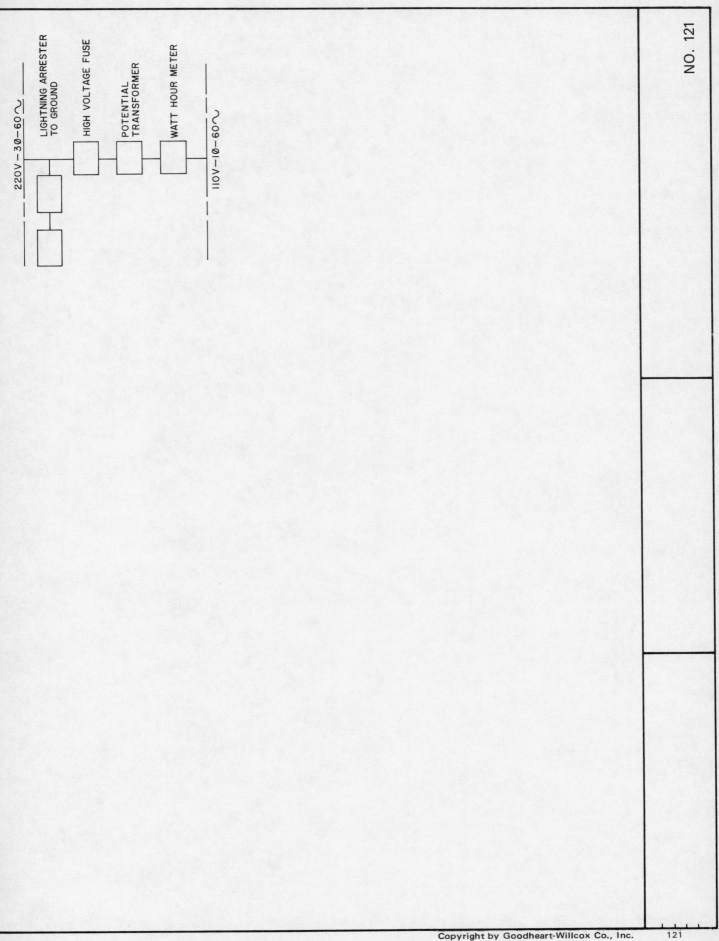

220V – 3∅ – 60 〜

LIGHTNING ARRESTER
TO GROUND

HIGH VOLTAGE FUSE

POTENTIAL
TRANSFORMER

WATT HOUR METER

110V – 1∅ – 60 〜

NO. 121

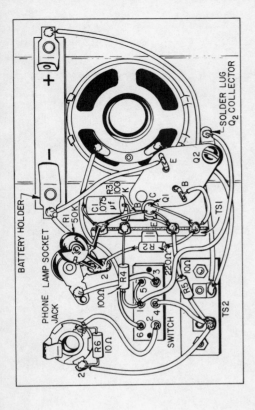

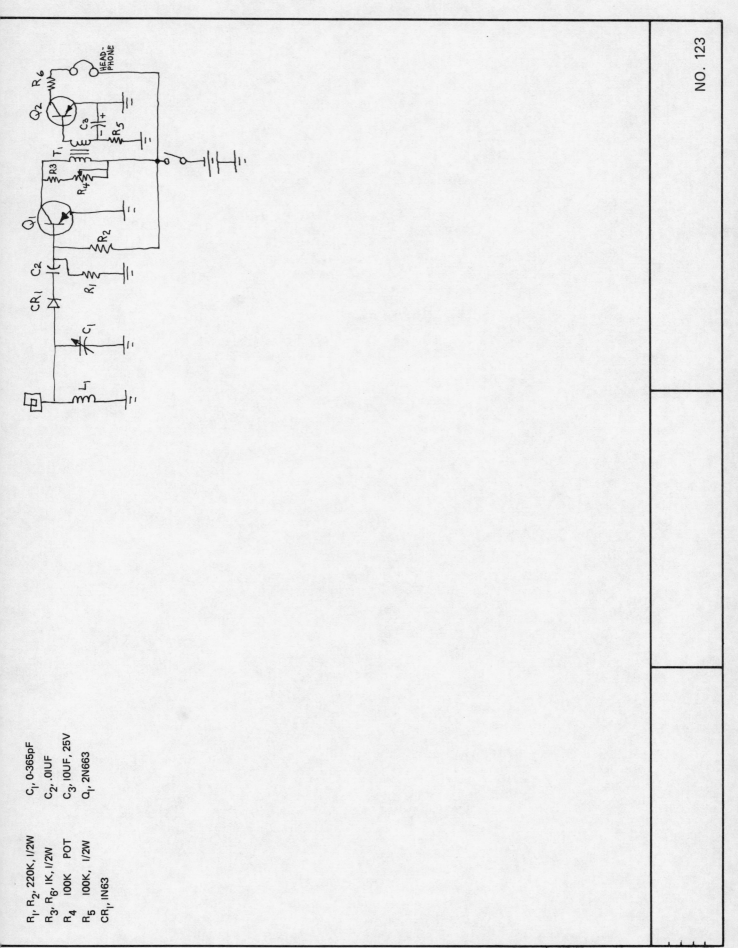

R₁, R₂ 220K, 1/2W

R₃, R₆, 1K, 1/2W

R₄ 100K POT

R₅ 100K, 1/2W

CR₁, IN63

C₁, 0-365pF

C₂ .01UF

C₃, 10UF, 25V

Q₁, 2N663

MAKE WORKING DRAWINGS, INCLUDING THE SPECIFICATION OF WELDS BY SYMBOLS AND MATERIALS LISTS FOR THE OBJECTS BELOW. USE A- OR B-SIZE DRAWING SHEETS.

PROBLEM NO. 124

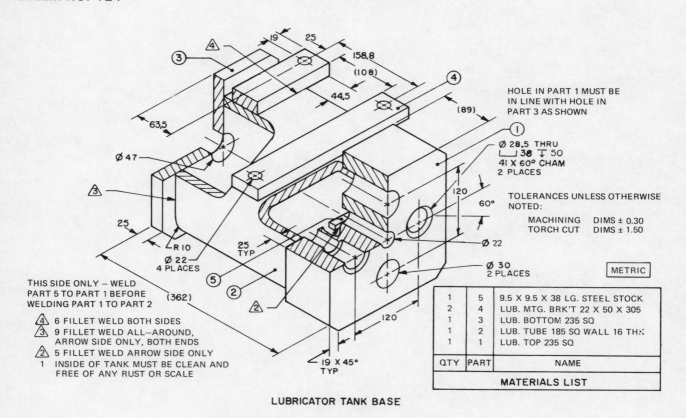

HOLE IN PART 1 MUST BE
IN LINE WITH HOLE IN
PART 3 AS SHOWN

Ø 28.5 THRU
⌴ 38 ⍗ 50
41 X 60° CHAM
2 PLACES

TOLERANCES UNLESS OTHERWISE
NOTED:

MACHINING   DIMS ± 0.30
TORCH CUT   DIMS ± 1.50

METRIC

THIS SIDE ONLY – WELD
PART 5 TO PART 1 BEFORE
WELDING PART 1 TO PART 2

△4 6 FILLET WELD BOTH SIDES
△3 9 FILLET WELD ALL–AROUND,
     ARROW SIDE ONLY, BOTH ENDS
△2 5 FILLET WELD ARROW SIDE ONLY
 1   INSIDE OF TANK MUST BE CLEAN AND
     FREE OF ANY RUST OR SCALE

QTY	PART	NAME
1	5	9.5 X 9.5 X 38 LG. STEEL STOCK
2	4	LUB. MTG. BRK'T 22 X 50 X 305
1	3	LUB. BOTTOM 235 SQ
1	2	LUB. TUBE 185 SQ WALL 16 THK
1	1	LUB. TOP 235 SQ

MATERIALS LIST

LUBRICATOR TANK BASE

PROBLEM NO. 125

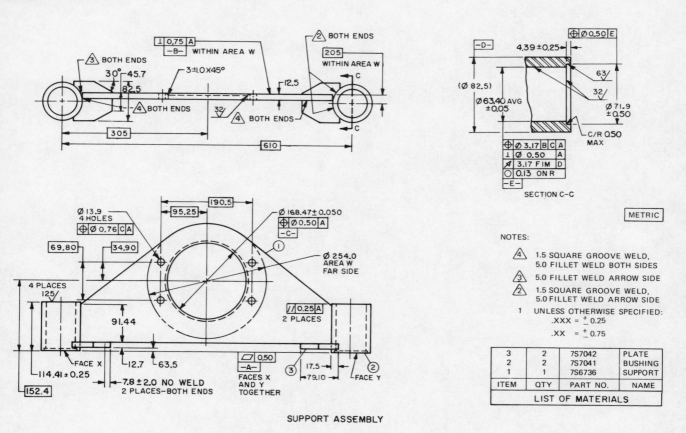

SECTION C-C

METRIC

NOTES:

△4 1.5 SQUARE GROOVE WELD,
     5.0 FILLET WELD BOTH SIDES
△3 5.0 FILLET WELD ARROW SIDE
△2 1.5 SQUARE GROOVE WELD,
     5.0 FILLET WELD ARROW SIDE
 1   UNLESS OTHERWISE SPECIFIED:
     .XXX = ± 0.25
     .XX = ± 0.75

ITEM	QTY	PART NO.	NAME
3	2	7S7042	PLATE
2	2	7S7041	BUSHING
1	1	7S6736	SUPPORT

LIST OF MATERIALS

SUPPORT ASSEMBLY

SELECT AN APPROPRIATE CONTOUR INTERVAL AND PLOT THE CONTOURS, AND NATURAL AND MAN-MADE FEATURES FOR THE MAP SHOWN. IN THE SPACE BELOW THE CONTOURED MAP DRAW A PROFILE MAP AT LINE 3. EXAGGERATE THE SCALE TO EMPHASIZE CHANGES IN ELEVATION. TITLE: TOPOGRAPHIC MAP WITH PROFILE.

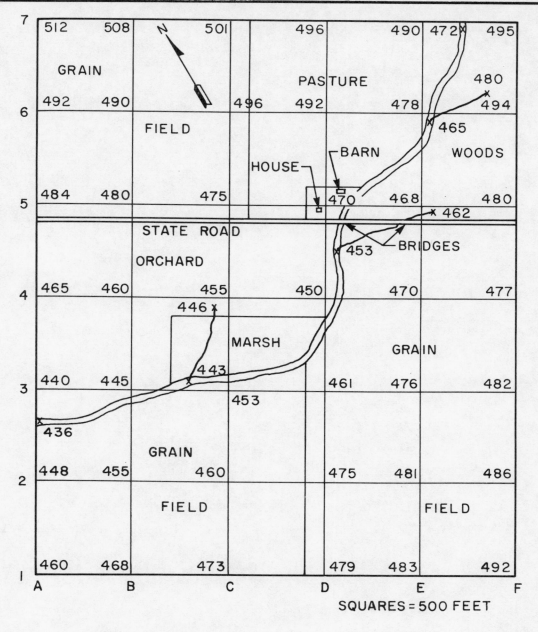

SQUARES = 500 FEET

NO. 126

COMPLETE THE MAP TRAVERSE BY PLOTTING THE DATA FOR STATIONS 2 AND 3. STATION NO. 2: RIGHT-DEFLECTION ANGLE = 75°;
DISTANCE FROM NO. 1 = 129 FEET. STATION NO. 3: RIGHT-DEFLECTION ANGLE = 138° 30'; DISTANCE FROM NO. 2 = 162.5 FEET.
CLOSE THE TRAVERSE AND INDICATE THE DIRECTION AND DISTANCE. SCALE : 1" = 30'.

STATION NO. 1

S 17° W
114 FEET

NO. 127

126

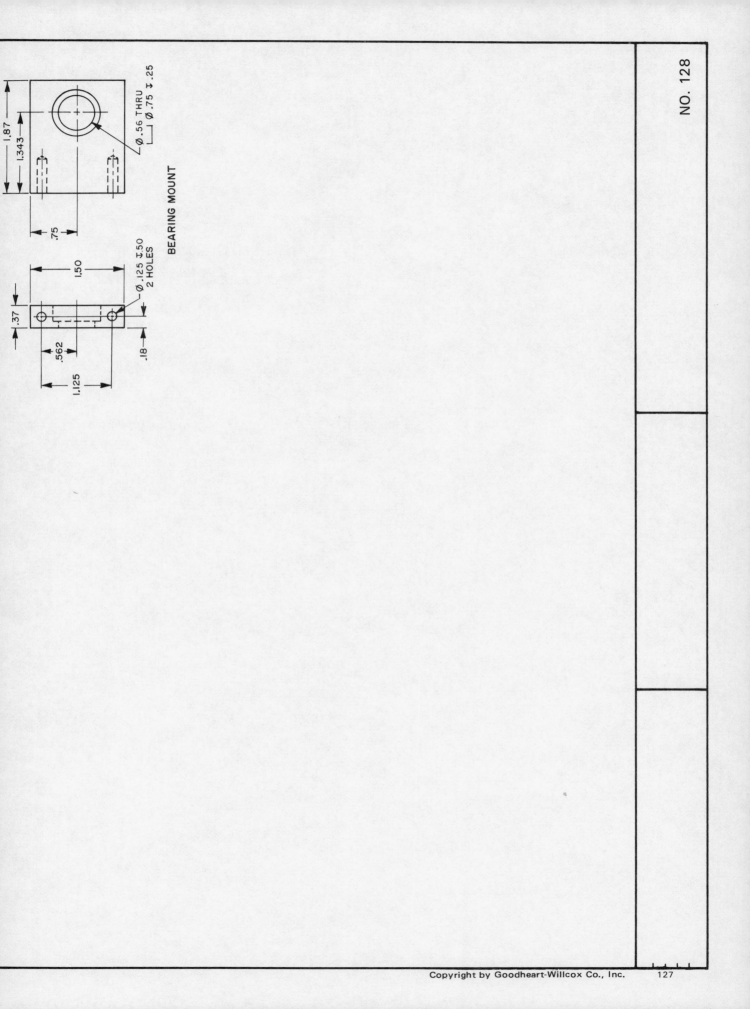

BEARING MOUNT

Ø .56 THRU
Ø .75 ⌴ .25

1.87
1.343
.75

1.50
Ø .125 ⌴ .50
2 HOLES
.37
.562
.18
1.125

.75

.38

4.00

.50

R .25
3 SLOTS

.75

.87

3.00

1.50

.50

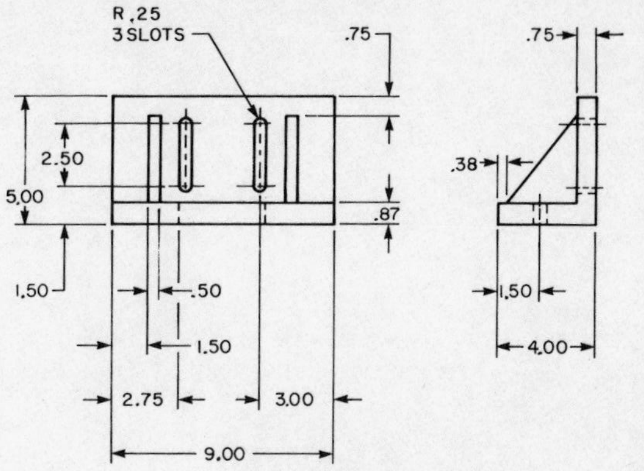

2.75

9.00

2.50

5.00

1.50

SLOTTED ANGLE PLATE

DATA. SUBSTITUTE ACTUAL DATES FOR YEARS.

MILLIONS OF WORKERS		PAST		PRESENT	FUTURE	
		10 YRS	5 YRS		5 YRS	10 YRS
SERVICE		24.8	27.2	34.0	47.6	59.7
GOODS		27.3	26.0	24.8	28.5	30.0

NO. 130

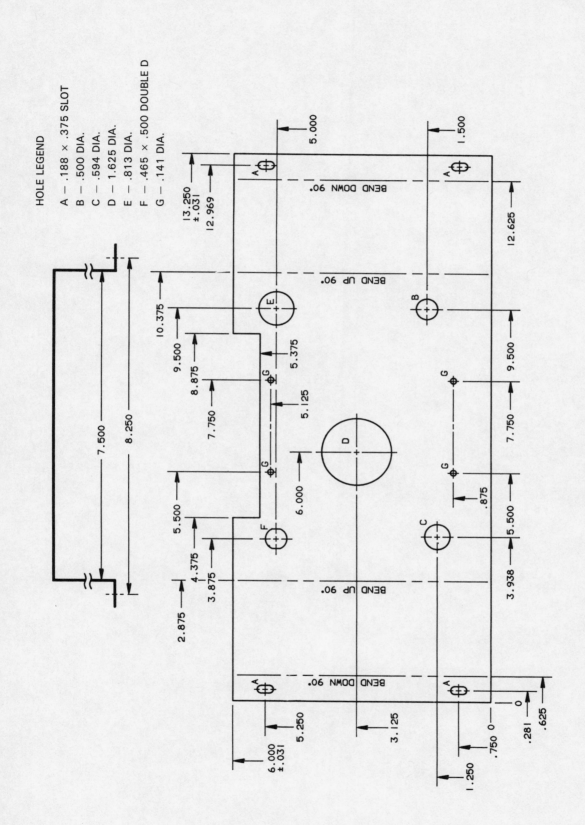

HOLE LEGEND

A — .188 × .375 SLOT
B — .500 DIA.
C — .594 DIA.
D — 1.625 DIA.
E — .813 DIA.
F — .465 × .500 DOUBLE D
G — .141 DIA.

COMPONENTS BRACKET

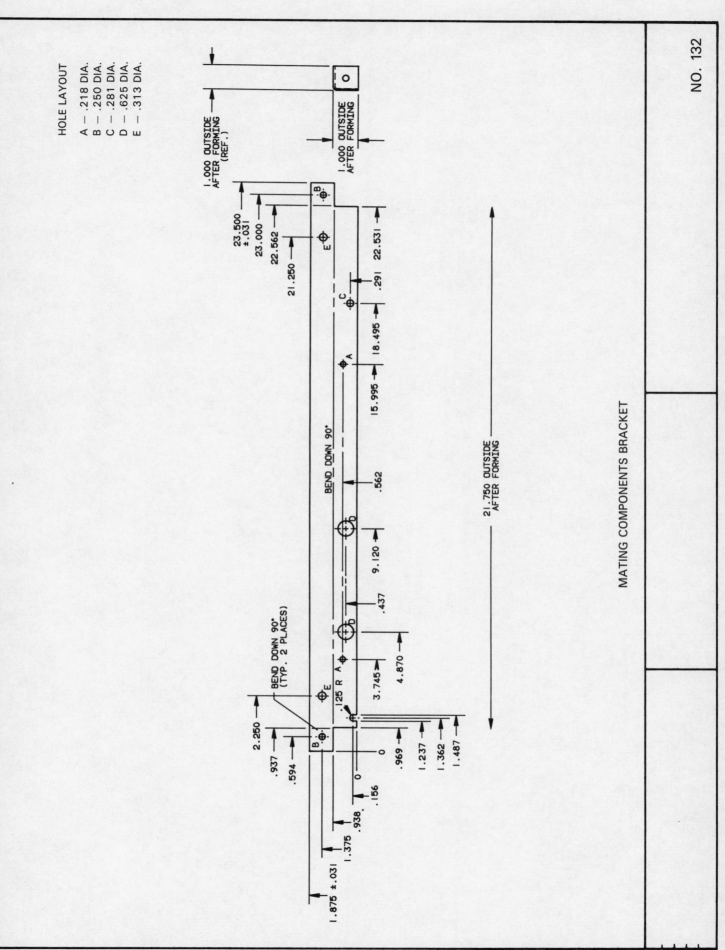

HOLE LAYOUT

A — .218 DIA.
B — .250 DIA.
C — .281 DIA.
D — .625 DIA.
E — .313 DIA.

MATING COMPONENTS BRACKET

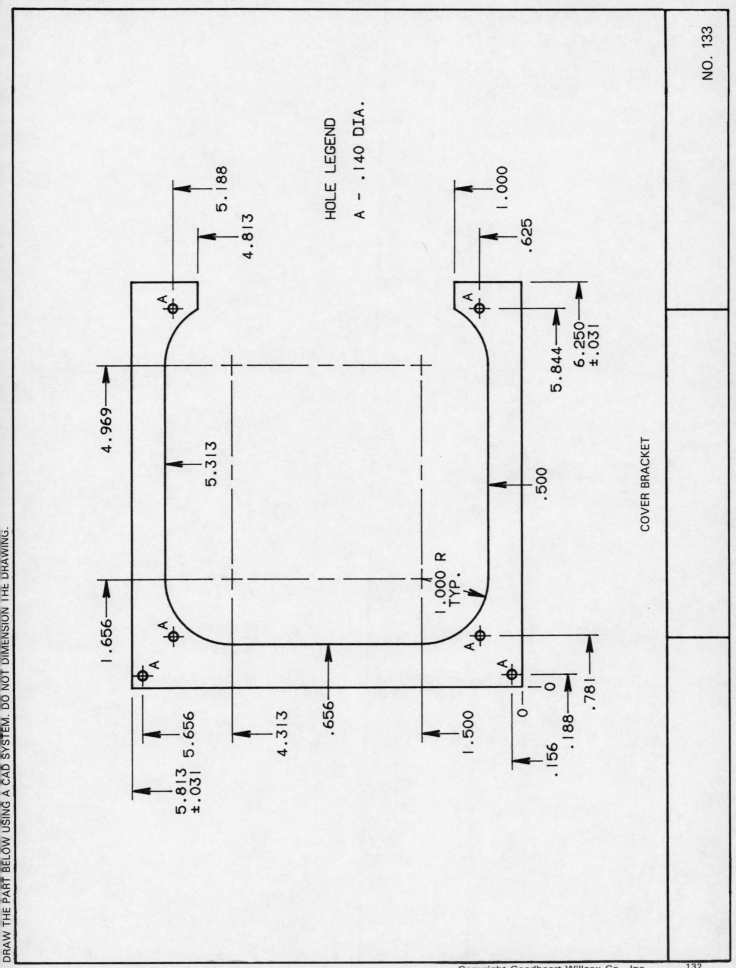

HOLE LEGEND

A — .140 DIA.

COVER BRACKET

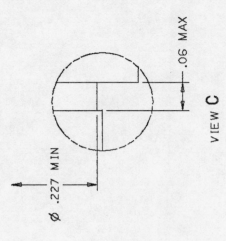

.06 MAX

VIEW C

Ø .227 MIN

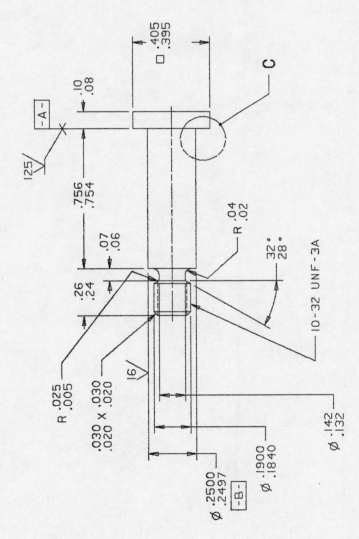

.405
.395

□

C

.10
.08

-A-

125

.756
.754

.07
.06

R .04
.02

.26
.24

32°
28°

10-32 UNF-3A

R .025
.005

.030 x .030
.020 x .020

16

Ø .142
.132

Ø .2500
.2497

-B-

Ø .1900
.1840

HANGER BOLT

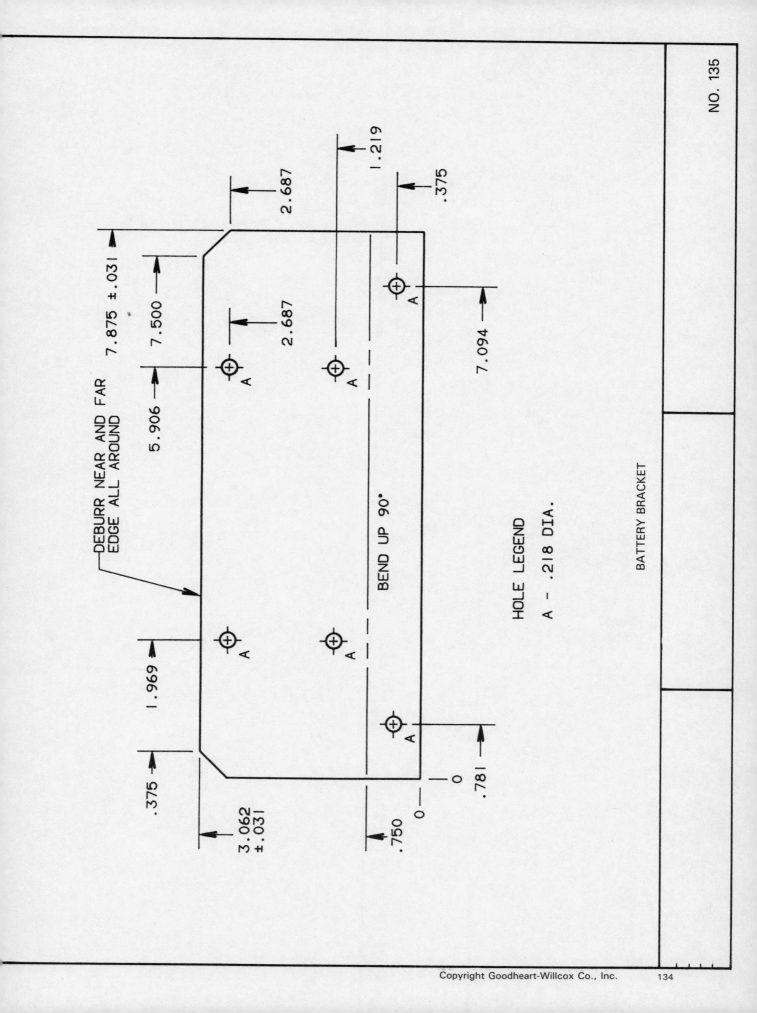

DEBURR NEAR AND FAR
EDGE ALL AROUND

7.875 ± .031

7.500

5.906

2.687

2.687

1.219

.375

7.094

A

A

A

BEND UP 90°

1.969

.375

3.062
± .031

.781

.750

0

0

A

A

HOLE LEGEND

A – .218 DIA.

BATTERY BRACKET

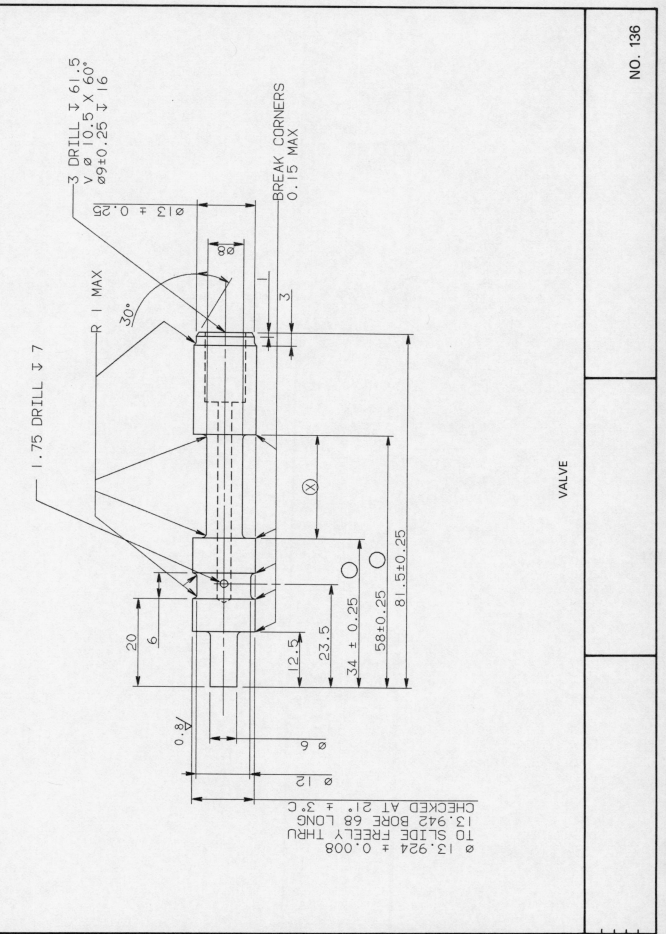

VALVE

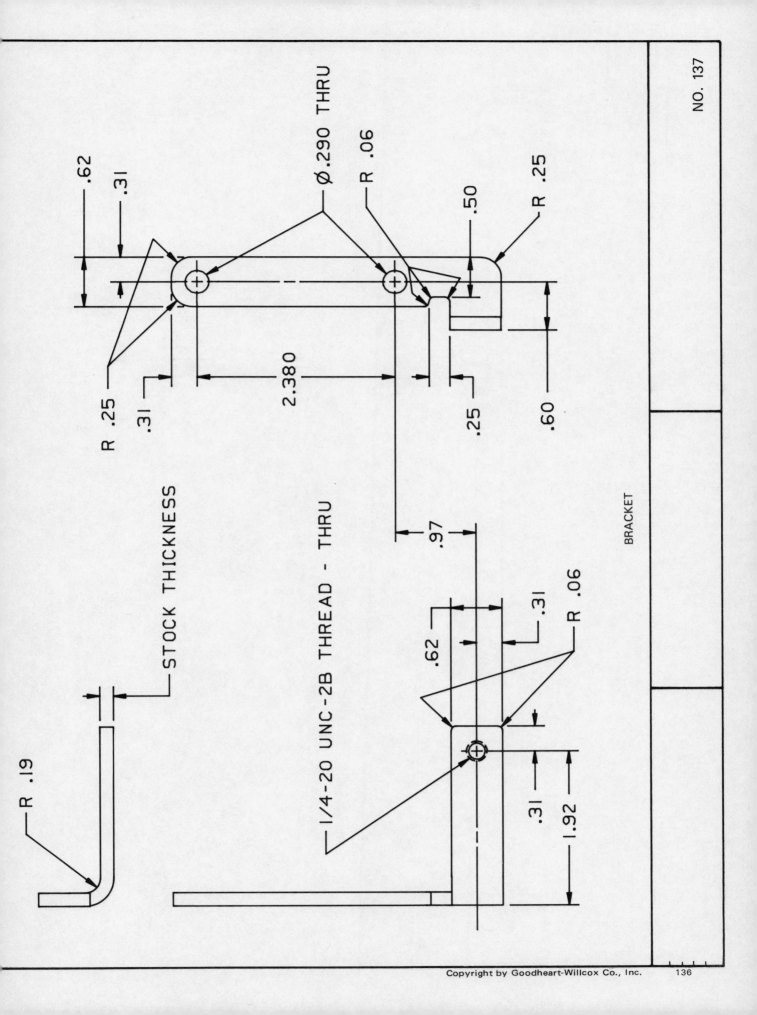

.62
.31
Ø.290 THRU
R .06
.50
R .25
R .25
.31
2.380
.25
.60
STOCK THICKNESS
1/4-20 UNC-2B THREAD - THRU
.97
.62
.31
R .06
.31
1.92
R .19

BRACKET

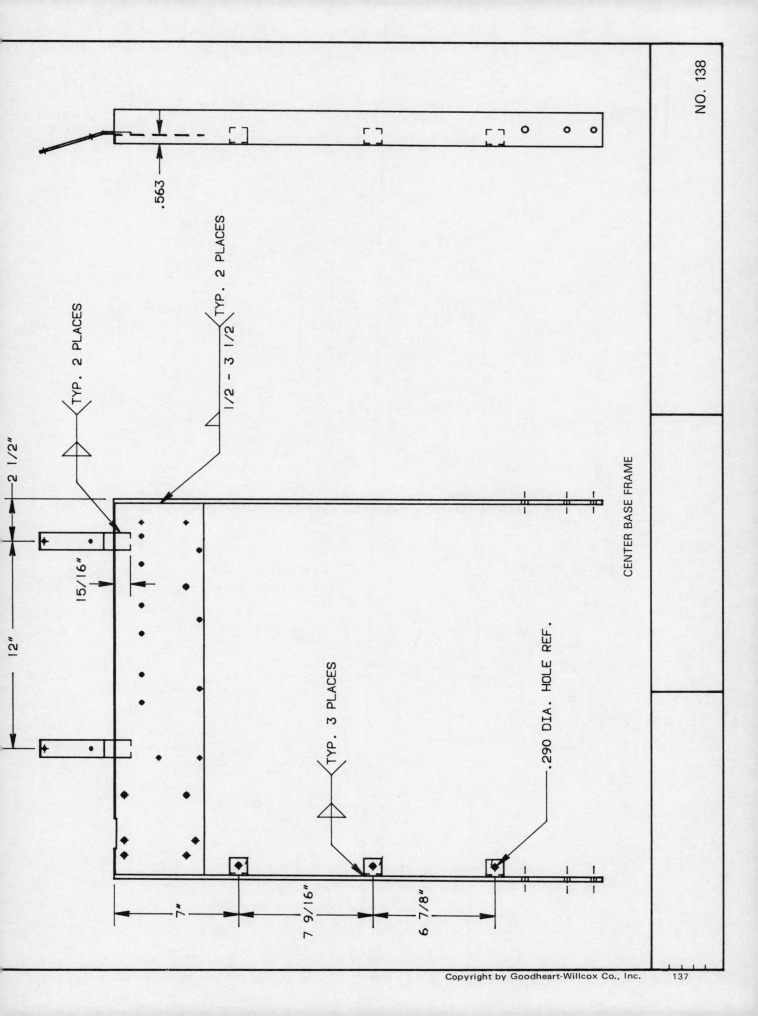

TYP. 2 PLACES

TYP. 2 PLACES

1/2 – 3 1/2

.563

2 1/2"

12"

15/16"

TYP. 3 PLACES

.290 DIA. HOLE REF.

CENTER BASE FRAME

7"

7 9/16"

6 7/8"

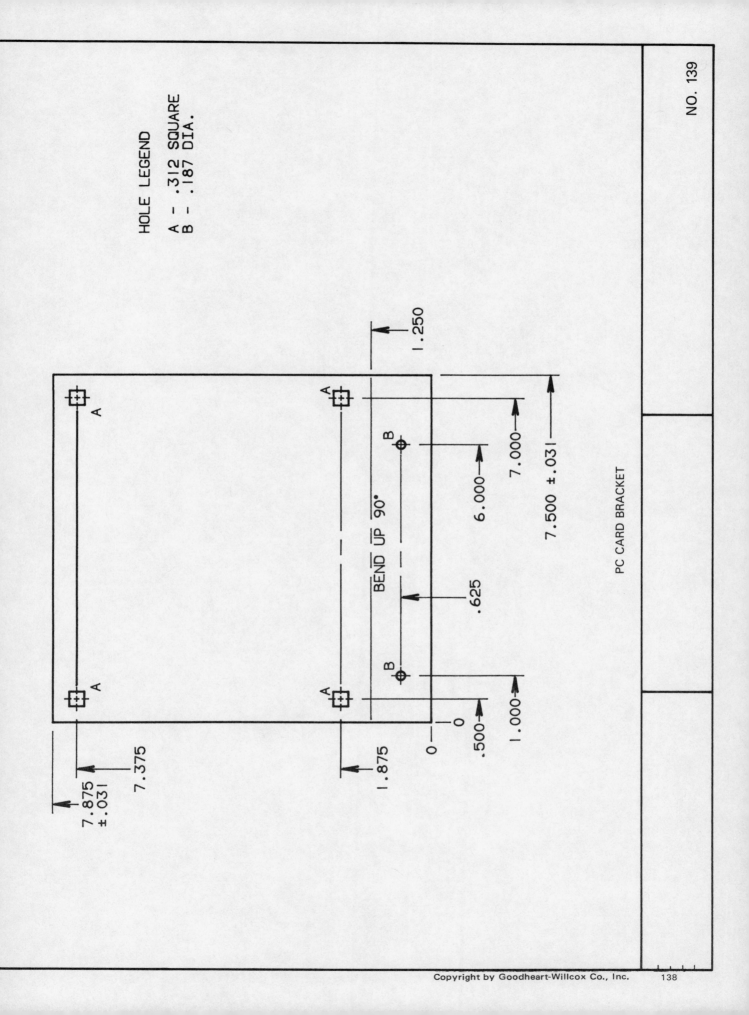

HOLE LEGEND

A – .312 SQUARE
B – .187 DIA.

BEND UP 90°

1.250

7.000

6.000

.625

7.500 ±.031

.500

1.000

1.875

0
0

7.375

7.875
±.031

PC CARD BRACKET

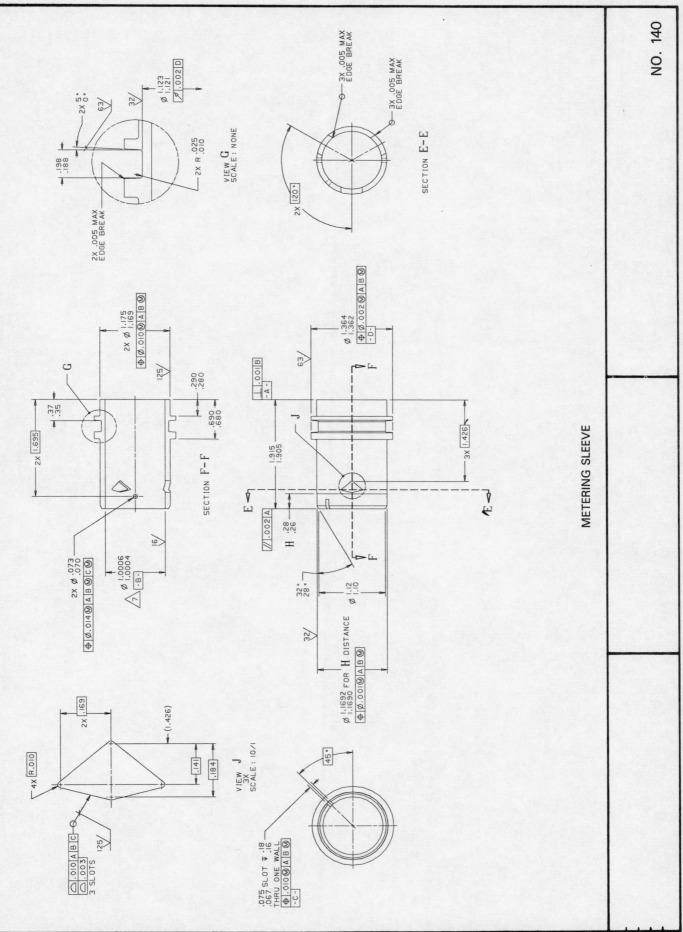

VIEW G
SCALE: NONE

SECTION E-E

SECTION F-F

VIEW J
3X
SCALE: 10/1

METERING SLEEVE

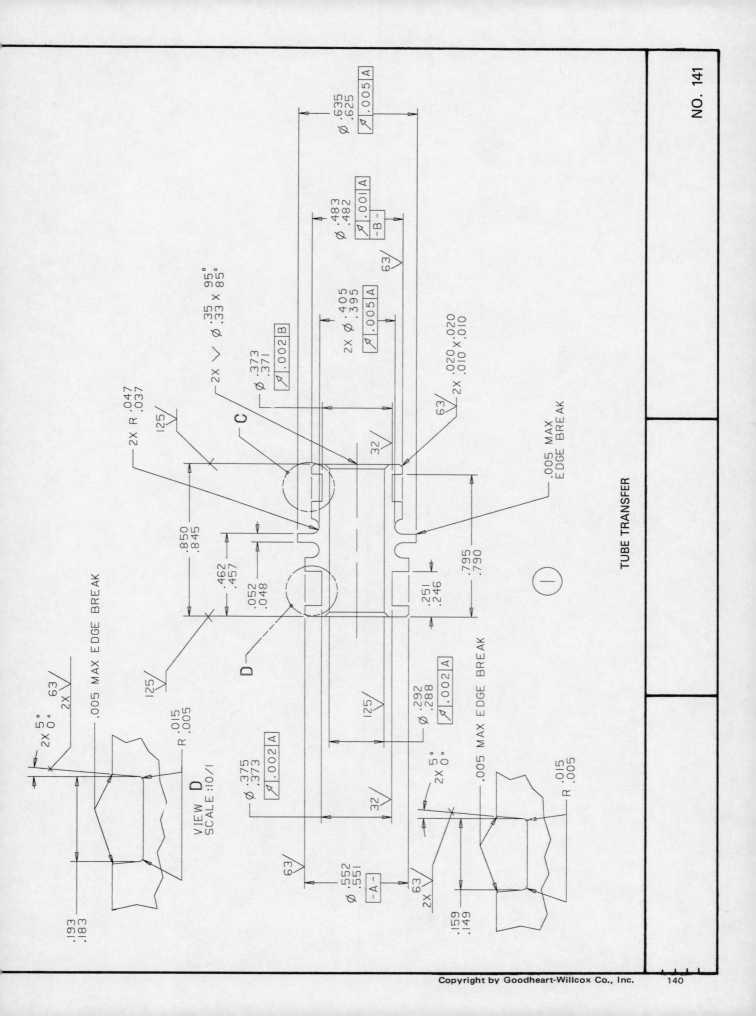

TUBE TRANSFER

VIEW D
SCALE :10/1

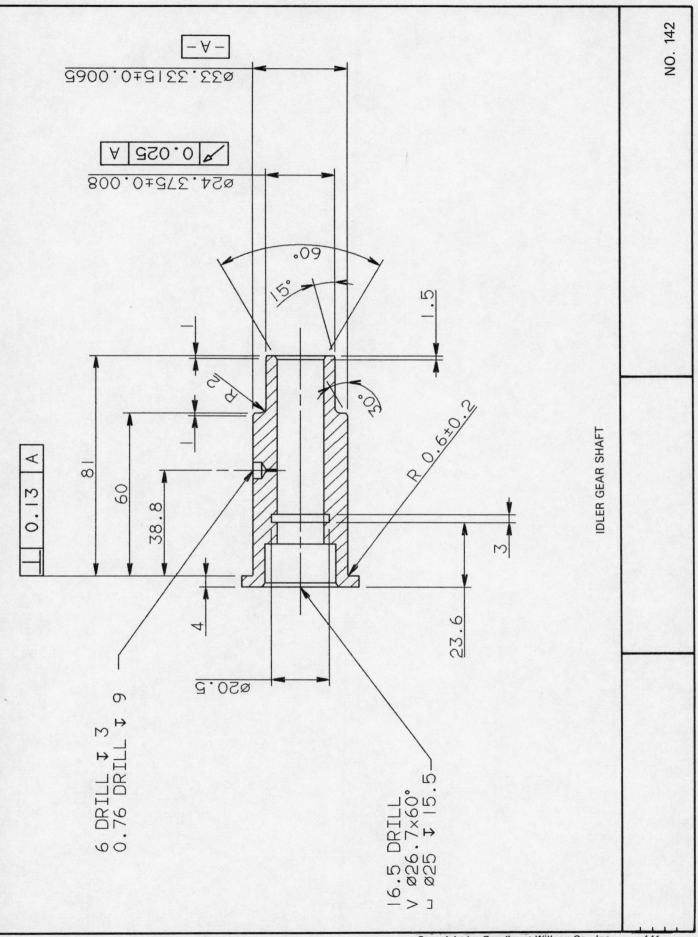

IDLER GEAR SHAFT

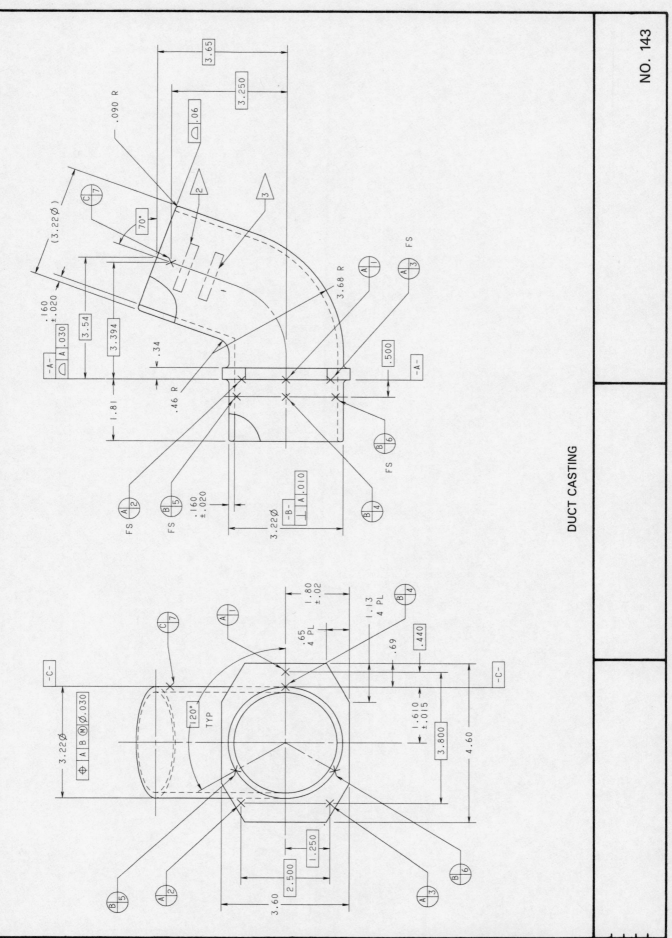

DUCT CASTING

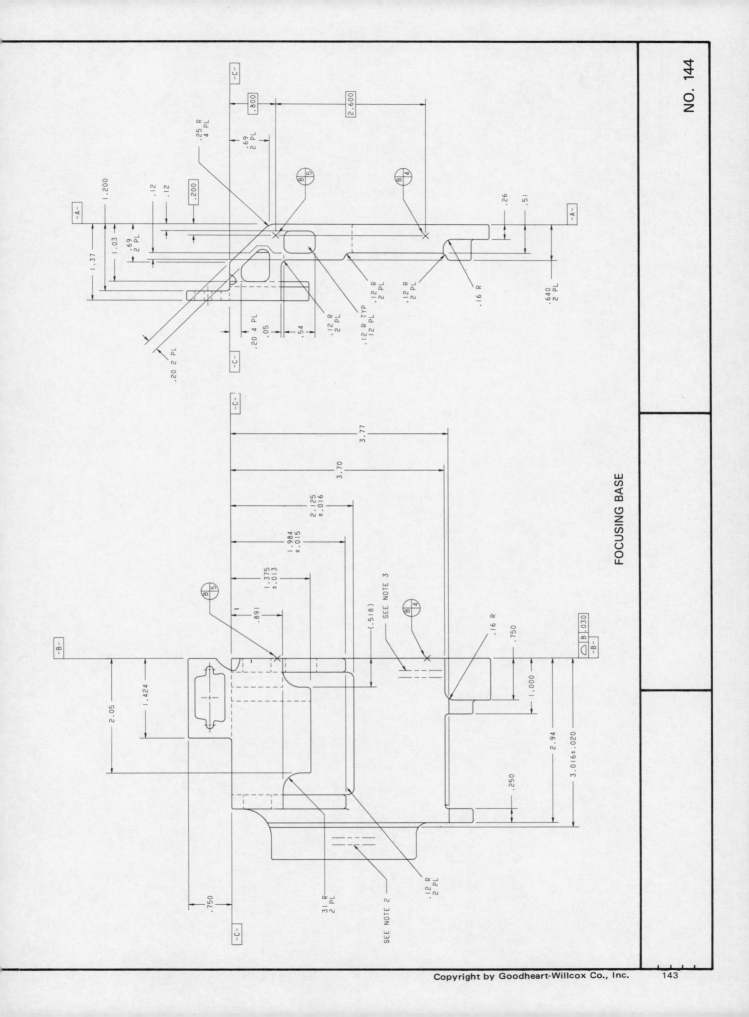

FOCUSING BASE